Secrets of
PLANT
PROPAGATION

Secrets of
PLANT
PROPAGATION

Starting Your Own Flowers, Vegetables, Fruits, Berries, Shrubs, Trees, and Houseplants

Lewis Hill

A Garden Way Publishing Book

Storey Communications, Inc.
"America's Garden Publisher"
Pownal, Vermont 05261

Acknowledgment

My thanks go to many friends who helped make this book possible, among them Dr. Norman Pellett and Dr. Leonard Perry of the University of Vermont Extension Service, and Steve Justis of the Vermont Department of Agriculture, as well as Greg Williams, Bart Hall-Beyer, and other nurserymen who have generously shared their expertise. I am grateful also to artist Elayne Sears, designer Andrea Gray, and most of all to my helpful and cooperative editor, Roger Griffith.

Design by Andrea Gray

Illustrations by Elayne Sears

Copyright 1985 by Storey Communications, Inc.

The name Garden Way Publishing has been licensed to Storey Communications, Inc. by Garden Way, Inc.

Ninth Printing, January 1991
Printed in the United States by Courier

Library of Congress Cataloging in Publication Data

Hill, Lewis, 1924—
  Secrets of plant propagation.

  Includes index.
  1. Plant propagation.   I. Title.
SB119.H55   1984   631.5′3   84-47788
ISBN 0-88266-371-2
ISBN 0-88266-370-4 (pbk.)

Contents

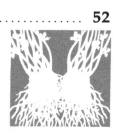
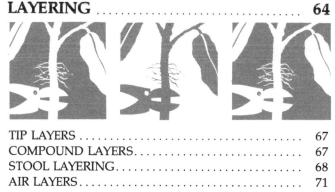

Chapter VI
GRAFTING 89

Chapter VII
BUD GRAFTING 104

Chapter VIII
TISSUE CULTURE 113

Part II
SPECIFICS OF PROPAGATION

Chapter IX
FRUITS & NUTS 119

Chapter X
TREES, SHRUBS & VINES 127

Chapter XI
HERBACEOUS PLANTS 142

Introduction

When I was growing up in the northern mountains of Vermont, most of the people I knew raised Holsteins, Jerseys, and Guernsey cattle. Although dairy farming was their primary business, they also had large gardens, and usually a small orchard of fruit trees and berries. None would have qualified as gardeners, as English horticulturists define the word, but they were as proud of their vegetable gardens as they were of their herds. Every row was straight, no weeds showed, and a less than perfect garden would have been as disgraceful to a proud Yankee as hanging out a dingy Monday wash would have been to his wife.

Putting in the garden on Memorial Day was a special occasion in our family and by the time I was six, I was allowed to drop in some of the larger seeds. A few years later I was even permitted a small patch of my own, and for many years, in spite of adult warnings, I could never resist digging down to see why the sprouts had not appeared by the third day. Perhaps because my childhood heroes were Peter Rabbit and Uncle Wiggily, it was especially fun to nibble the first radish and tender greens. Later I tested the new peas and young carrots, the latter being hastily, and not too carefully, wiped on the grass before consumption. I especially enjoyed the thick sandwiches we kids used to make by squashing wild red strawberries between lettuce leaves. Gardening was always more fun than barn chores.

Although my fondness for fresh fruits and vegetables influenced my interest in plants, I had no intention of becoming a gardener. Instead, like many of the teenagers of my era, I had carefully planned a Horatio Alger career in the city, but fate intervened. When I was in high school, a man from Scarsdale, New York, who spent his summers in my town, asked if I would send him a dozen raspberry plants from our patch at the proper planting time in the spring. I nervously dug up some sucker plants, cut off the tops, wrapped them up to look as much as possible like the nursery stock we occasionally bought from Ward's, and took them to the post office. They promptly refused to accept the plants without a shipping permit from the state nursery inspector. Since I had promised the plants, I wrote to the Department of Forestry. Within a few days the elderly inspector arrived, looked over my tiny patch, and gave me a permit and lots of advice about starting a nursery. Since I had no notion of going into the plant business, I paid little attention.

Delighted with the Vermont plants, my Scarsdale friend asked if I could send him a variety of small native evergreens. Since I already had the permit, I obliged, and he started recommending me to his friends. Soon many people were asking for a variety of plants, many of which were unfamiliar to me. I started a plant collection in the backyard, and began doing lots of reading.

Gradually I realized that propagating plants was not only profitable, but it was more fun than mere gardening. Our little nursery grew bigger each year. Some of the local horticulturists showed me many of the traditional ways of starting plants, and the nursery inspector, who also taught at the university, brought new suggestions at each visit. He also supplied me with countless extension bulletins, most of them written by himself.

My early attempts at propagation were filled with many failures because, in spite of the help I was getting, there was still a lot of necessary information I couldn't find. Each success was encouraging, however, and in the long winter, I tried to devise methods of correcting mistakes of the previous season.

I hope that this book will make available to you the necessary techniques for propagating your own plants and that it will help you avoid the mistakes and disappointments that I had as a novice in the forties and fifties. In the first section I have explained the step-by step process of each propagation method and, equally important, told you how to care for your newly started plants for the first critical weeks of their life. The second section of the book divides plants into three: (1) fruits and nuts, (2) woody trees and shrubs, and (3) perennial herbaceous plants. Propagating methods are recommended for each plant listed.

Whether you decide to start only a few plants as a hobby, or plan to open a small nursery or garden center, you are embarking on a truly exciting and satisfying project. I wish you well.

Part I

PROPAGATION METHODS

THE HOME NURSERY

By planting a seed, rooting a cutting, attaching a graft, and the other myriad activities of plant propagation, you "hold infinity in the palm of your hand," as William Blake wrote in his *Auguries of Innocence.* You are involved in the process of creating new life—that awesome natural force that is far beyond man's understanding, but is an integral part of life. Perhaps that explains the underlying pleasure and joy we have in starting plants.

Most of us who become intrigued with plant propagation initially are interested only in the final product, however. We plant seeds for lettuce, raspberry bushes for their fruit, acorns for future shade, or geranium slips to brighten up the window box. Later, as we get more caught up in the process, we discover how challenging and exciting the hobby of starting new plants can be.

Plants can be propagated either sexually or asexually. In sexual reproduction, a completely new individual is produced from each seed, just as when an animal baby is born, and the parents will eventually die. In asexual propagation, however, a part of the parent is removed and used to start a new plant, so in theory the parent lives on indefinitely. A grape that grew back in Biblical times may still be living today by this continual cloning process, and still be almost identical to the original vine.

Sexual Propagation

The sexual method to increase plants is by far the most widespread. Nearly all the vegetation of the world is produced by seeds, including grains, forest trees, vegetables, and grass. Seeds are used not only for raising new plants that are duplicates of the plants they came from, but are also used in breeding to create entirely different specimens. Experimenters use chemicals or other laboratory techniques to alter the cell makeup of the seed, for the purpose of developing a variety quite unlike either parent.

Asexual Propagation

Because many seeds do not reproduce exactly the characteristics of their parents, asexual propagation, sometimes called vegetative propagation, is used to produce a new plant that is like the parent. Plant cells are programmed with the genetic ability to reproduce a duplicate plant. The method of regeneration may range from starting a houseplant slip in a glass of water, to extremely complicated, high-technology systems. Whatever the technique, in asexual propagation a piece is taken from one plant and a new plant is started from it by division, cutting, layering, or grafting. By tissue culture, the most recent space age method, plants are started in test tubes under absolutely sterile conditions.

Use of Two Methods

Plant nurseries use asexual propagation more frequently than they use sexual because they specialize in starting varieties of trees, shrubs, and perennials that do not "come true" when started from seed. Seeds from such plants are likely to produce new plants that have characteristics of one or more of their ancestors. A McIntosh apple seed, for example, would not produce a McIntosh tree, but, more likely, a tree bearing quite different fruit. Fruit trees, small fruits, grapes, many perennials, and most named varieties of ornamental shrubs and trees are not propagated from seed for this reason.

Most home gardeners, on the other hand, tend to grow their plants by the sexual method, and sow annual flower and vegetable seeds every spring. They buy their flowering shrubs, flowering trees, fruit plants, perennials, herbs, and evergreens already started.

Select the Best Method

Although plants can be reproduced by several methods, they usually can be propagated best by a particular one. Mountain ash, for example, are nearly always grown from seed, as are oaks and maples. Named varieties of tea roses and fruit trees are usually started from some kind of graft, hedge plants reproduce quite easily from hardwood cuttings, and most ornamental shrubs start best from softwood cuttings.

Numerous plants can be propagated in several ways. Trees like poplar and willow trees start easily from hardwood or softwood cuttings, root cuttings, suckers, and seeds. Orchids, on the other hand, are difficult to propagate by any ordinary process and now are being reproduced almost exclusively by tissue culture.

Do It at Home

The "green thumb" theory and mystique are dispelled when you understand the natural laws that govern propagation. If you want, you can do at home most of what is done in a commercial nursery, from starting a single bush or perennial to growing hundreds of little fruit trees or evergreens. One friend of ours has an elaborate setup for propagating all kinds of plants. He operates grow lights in his basement, hotbeds, a small greenhouse, a mist area for starting cuttings, and outdoor beds for his grafts, seedlings, root cuttings, and transplants.

Although his hobby is limited to evenings and weekends, since he has a regular Monday to Friday job, he nevertheless produces a tremendous amount of plant material. His own property is full of fruit trees, bushes, and flowering plants, and he has saturated his friends with nursery stock. Because he can't bear to throw away his fledglings, he takes loads to farmers' markets, and donates more to community auctions and sales.

Few home nurseries are as involved or as productive as his, and it is easy to start all the plants you want with a far simpler setup.

EQUIPMENT

Although many seeds, cuttings, and grafts can be started outdoors successfully, gardeners who live in cold or very warm areas need a climate for starting plants that is more hospitable than that which Nature provides. Most plants can be propagated successfully only when environmental conditions provide the proper temperature, humidity, and light for good root and top development. Satisfactory conditions may range from a glass of water placed on a kitchen shelf for rooting a slip of ivy, to a well-equipped greenhouse where temperature, ventilation, shade, watering, and fertilizing are controlled automatically.

This following list of suggested equipment for small-scale propagation includes some things that I consider essential if you intend to propagate a wide range of plants, and some that I recommend for the serious gardener as desirable, but not absolutely necessary. In plant propagation, as in photography or woodworking, you can become as involved as you want. It can be less expensive than many hobbies, because if you wish, you can build a great deal of your own equipment. Or you can buy it at hardware stores, garden centers, or from the nursery supply catalogs listed in the appendix at the end of this book.

Knives

Your cutting equipment should be high quality because cuttings and grafts require smooth, clean cuts. A high grade pocket knife is adequate for small propagating jobs, but you may want to consider one of the knives made especially for budding or grafting. They do the job better and are easier to keep sharp. Some propagators use razor blades, but I find them hard to handle and whenever I use them, I require a lot of Band-Aids.

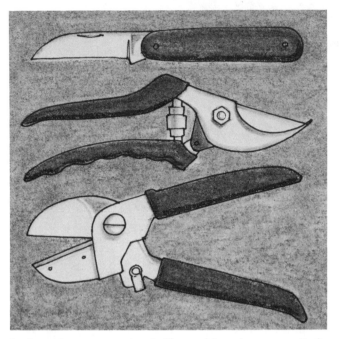

Basic tools are a pocket knife and hand pruners. Both should be kept clean and sharp.

Hand Pruners

Hand pruners come in many styles and prices. Although you may not need the very best, the cheapest should not be considered either. There are many good clippers in the in-between price range. Dull, poor quality clippers crush the wood rather than cutting it smoothly. Pruners come in two different styles, the shear type and the anvil one. Some like the shear type because they can make a closer cut. Others choose the anvil ones, feeling they are less likely to twist out of shape under a heavy cut. Whichever you select, keep them sharp, clean, and adjusted properly.

The Notebook

As time goes on, a notebook will be indispensable for many reasons. It is easy to forget the names of varieties, dates, and other pertinent data. If you experiment with various chemicals or methods, you will want to know in subsequent years what worked and what didn't. It will help you to plan each gardening year, remind you of work to be done each season, which supplies to order, and in what amounts. It should show maps of your gardens and orchards, as well.

Entries in your notebook might look like these:

April 3: Grafted three potted apple tree seedlings. (A) McIntosh, (B) Baldwin, (C) Delicious.

ESSENTIAL EQUIPMENT AND MATERIALS

Equipment

Hand pruners	Bulb sprinkler
Knives and a sharpening stone	Hand sprayer
	Soil sifting screen
Trowel	Flats, trays, and pots of various sizes
Spade	
Dibble	Stakes
Sprinkler can	Grow lights
Hose and fine spray nozzle	Thermometer

Materials

Vermiculite	Insecticides
Perlite	Notebook
Peat moss	Labels
Soil	Marking pens and pencils
Sand	
Pro-Mix or other artificial soil mix	Plastic wrap
	Clear plastic bags
Peat pots or fiber cubes	Twine
	Budding strips
Fertilizers	Tree dressing
Rooting chemicals	Grafting wax or tape
Fungicides	

Equipment Less Essential but Handy to Have

Potting shed	Grafting wax melter
Hotbed	Soil shredder
Cold frame	Soil test kit
Greenhouse	Garden cart or wheelbarrow
Outdoor seed and transplant beds	Mist system

Grafts from Hawkins's trees next door. Used wax on (A), tape on (B), and Saran-wrap on (C).

Transplanted tomatoes and peppers to large pots. Potting soil from tub A. Mixed MagAmp fertilizer in soil. Watered with manure tea.

Checked orchard for winter damage, cleaned up prunings. Pruned off broken canes on black raspberries.

April 5: Cold, rainy day. Seedlings in greenhouse showing signs of distress. Cancelled watering for the day, and increased heat. Sprayed with Subdue.

PROPAGATION MEDIA

Any material in which seeds are planted, cuttings inserted, or plants are set, is called the "medium." Nature's own soil is, of course, the most common, but there are many others as well. Some are organic, such as peat moss and compost, and others, such as vermiculite and perlite, are inorganic. To be successful for starting plants, a medium should have a good exchange ability; that is, it should be able to absorb and release to the plants the amounts of moisture and fertilizer necessary for them to start and grow. A good propagation medium should also be sterile — free from soil diseases and weed seeds. Since different kinds of seedlings and cuttings vary in their requirements, no one medium is considered perfect for all plants and growing conditions.

Artificial Soil Mixes

Most gardeners have the best luck starting seeds in a sterile artificial soil such as Jiffy-Mix, Pro-Mix, Metro-Mix, or others which are available at garden centers and farm stores. These consist of various blends of vermiculite, perlite, peat moss, small amounts of fertilizer, and sometimes a wetting agent. Of course, it is possible to mix your own, but you will save time, mess, and dust by using one that is already mixed and sterilized.

When buying an artificial soil mix, don't confuse it with the potting soils that are often sold in garden centers and department stores. Potting soils are fine for repotting house plants and transplanting seedlings, but they are too heavy for starting seeds or rooting cuttings, and often they are not completely sterile.

Compost

Composted leaves and hardwood bark were once common media for starting seeds. They are relatively free from weeds, and seeds grow well in them because they are fertile and rich in good soil bacteria. Compost is easily infected with damping-off virus, however, so it should be sterilized before it is used for starting seeds.

Compost is very useful when mixed with less absorbent ingredients in media used for rooting cuttings, or for transplanting seedlings that are well started. Composted manure and composted garden wastes were once included in many seed starting mixes also, as was dried manure, but because of the odor, the favorable environment for breeding flies, and the difficulty of keeping them sterile, none of them is recommended.

Peat Moss

Peat moss is seldom used alone, either for starting seeds or rooting cuttings, because its sponge-like quality hinders good root development. It is an excellent addition to other media, however, since it increases absorbency. Its very high acidity makes it especially useful for starting acid-loving plants such as azaleas and blueberries. Because it is lumpy, peat moss should be sifted before it is mixed with other ingredients.

Perlite

Perlite is a volcanic glass that has been ground, and, in contrast to peat moss, it has very little ability to absorb and release fertilizer and moisture. For good exchange, therefore, it is essential to mix peat, vermiculite, composted bark, or some other material with it. I especially like the fine grade of perlite for covering newly planted seeds indoors, because it dries out quickly after

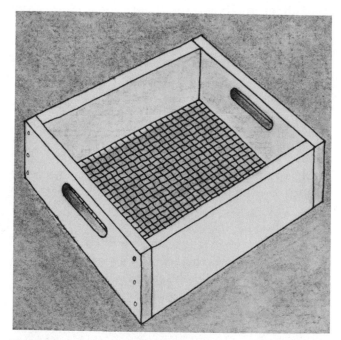

This sifting box is easy to make and is useful when preparing a fine compost mix for transplanting seedlings.

watering and keeps the surface of the planting medium dry, thus inhibiting disease. It is also an excellent soil lightener, to provide aeration and drainage.

I have found coarse perlite useful as a medium for starting the cuttings of certain plants that develop roots slowly. When placed under mist, their stems tend to rot before rooting begins in most media, but coarse perlite provides good drainage. The stems stay moist enough to root, but do not become wet enough to rot.

Sand

Sand is an old favorite for rooting cuttings. It is also used to help provide drainage and aeration in mixes that include peat moss, soil, and compost.

HOW TO STERILIZE SOIL OR OTHER MEDIA

Mix soil thoroughly before sterilizing, and moisten it slightly but uniformly. Small amounts can be sterilized in a home oven. A favorite old-time method was to put a metal kettle filled with moist soil into the oven with the potatoes being baked for dinner. When the potatoes were done, the soil was sterilized, but the house smelled terrible. Potatoes aren't necessary, however. Spread out a gallon or two of soil in shallow metal pans and bake it for a half-hour at 200 degrees F. to kill most harmful organisms.

Commercial machines inject hot steam into the soil through perforated pipes buried in it, and the temperature is kept at 180 degrees F. (82 degrees C.) for one-half hour. Some prefer using a slightly lower temperature for a bit longer, believing it doesn't kill off as many beneficial organisms in the soil.

Chemical Fumigating

Although heat is effective, chemicals are now used more frequently to fumigate soil. A good sterilant should control the harmful fungi, weed seed, and roots, without wrecking the soil structure.

Vapam, a favorite of both commercial and home gardeners, is useful for sterilizing potting soil, starting mixes, and outdoor seed beds. If you use it, follow the directions on the package carefully, because the sterilization will be ineffective and even dangerous if done carelessly. The liquid becomes a gas in the soil, and this gas should be completely released before the soil is used for planting or moved into a greenhouse. Otherwise, escaping gas may kill all the plants in or near it.

Poison gases such as methyl bromide, chloropicrin (tear gas), and other chemicals such as formaldehyde are used by commercial operators, but are not recommended for home gardeners.

If the soil is to be used for transplants or vegetable plants and rugged annual flowers, which are not susceptible to soil diseases, the only precaution that may be necessary is to drench the soil with a fungicide such as benomyl, rather than to sterilize it.

Because sterile soil can easily become reinfected, your hands, all trowels or other tools, and used flats and pots should be sterilized with a full-strength solution of Clorox or a similar bleach.

When preparing soil mixes, place all of the ingredients in a container.

Stir well, then moisten until soil is damp, and mix well again.

Place soil in flats or other containers. You're ready to plant seeds.

Clean, sharp river sand which is not too fine and contains no soil or other materials is the best kind to use for this purpose. Although it is usually free from harmful organisms, it should be sterilized before being used as a starting medium.

Soil

If soil is used for starting seeds or cuttings it must be sterilized first. Soil taken from gardens and other cultivated areas is especially likely to be loaded with potentially harmful soil organisms and weeds, but even virgin soil dug from a hardwood forest should never be used for starting seeds without sterilization.

Until quite recently sterilized soil mixed with sand and peat moss was used exclusively for starting seeds by nearly all commercial greenhouses. Although artificial soil mixes are now more popular for seed germination, sterilized sandy-loam soil is still used as a potting medium for transplants. Outdoor beds for starting seeds of perennials, trees, and shrubs are often made of a mixture of soil, sand, and peat or compost. This mixture can also be used satisfactorily as a seed starting medium for some of the less delicate plants such as the cabbage family, lettuce, corn, and beets.

Vermiculite

Vermiculite is exploded mica, the same material that is often used in home insulation. The horticultural grade is best for gardeners, but if only a coarse, home insulation grade is available, you can grind it finer in an old coffee grinder or blender. Small amounts can even be rubbed fine in your hands. Vermiculite is very useful in the home nursery, not only as a sterile medium for starting seeds and cuttings, but also for lightening soil mixes.

CONTAINERS

Some garden supply houses sell small peat pellets for starting seeds, or peat pots that you may fill with a soil mix. Plant two or three seeds in each pellet or pot, making sure they are covered with the peat or mix, and water. (After watering, the pellet swells to many times its original size.) When the seedlings are about a half-inch high, the extra ones should be pinched off, leaving only one single strong plant in each.

Because the roots grow through the peat, the whole unit can be transplanted intact into a larger pot or the ground and there is very little transplant shock. We use these for starting zucchini, cucumbers, squash, and other things that are not easy to transplant from flats.

Fiber, clay, and plastic pots are available from nursery supply houses. (See Appendix.) They may be round or square, and range from a bit

When starting seeds in peat pellets, first plant two or three seeds.

Cover them and water. Later clip off all but the strongest seedling.

Plant entire peat pellet. Roots will quickly grow through peat.

There are containers for a variety of needs, from large flats to small peat pellets.

over an inch in diameter to giant containers that will hold large trees. Hard and soft plastic pots are the most popular. Although the life expectancy of the plastic varies according to the quality of the material, as a rule, soft plastic pots last from two to four years, the hard plastic pots, three years to eight or more, and the fiber pots last only one or two years.

Some gardeners prefer the clay pots because the porous clay sides provide aeration for the soil. Clay pots are very heavy, fragile, and need more frequent watering. They are also much more difficult to wash and sterilize than plastic pots.

Flats also come in many sizes and materials. They range from a tiny unit useful for only two to four small plants to large trays which may be used for starting seeds, or for holding small flats or pots. They are made of hard plastic, metal, or fiber. Most growers prefer the hard plastic for starting seeds because they can be washed, sterilized, and reused. Either the hard plastic or fiber flats are used for selling transplants.

There is no need to buy expensive pots or flats, if you need only a few. The plastic tubs used for butter, margarine, cheese, yogurt, and ice cream make excellent containers. We often recycle styrofoam cups, which are excellent for small herb and flower plants. Punch drainage holes in each container on the sides close to the bottom of the pot rather than on the bottom. Holes on the bottoms of flats and pots often get blocked, especially if they are placed on a bench covered with plastic.

ARTIFICIAL CLIMATES

Because newly formed plants are so fragile, the constantly changing climate of the outdoors is not an ideal place for most of them to start life. Until they have grown a good root system and an adequate leaf surface, small seedlings and new rooted cuttings need some kind of protection. Although the seeds of many native plants and most weeds sprout under any condition, most seeds and cuttings are delicate and need special light, temperature, and humidity if they are to thrive. The most common methods of climate control used by home gardeners to start seeds and cuttings are:

Grow Lights

We have found that seeds germinate and grow best under grow lights because they provide ideal light and heat conditions. By leaving the lights on for twenty-four hours a day, initially, the seeds sprout rapidly and grow speedily. The new plants, consequently, are healthy, and usually resist damping-off diseases.

Ordinary incandescent light bulbs can be used for starting seeds, or you may use heat lights, fluorescent bulbs, or the large lamps used for outdoor security lights. The pink-colored fluorescent bulbs made especially for growing plants work best, however, because they emit the light rays most beneficial to plants.

Many garden supply houses sell grow-light bulbs and fixtures which you can mount over a bench or shelf. To start large numbers of plant or herbaceous cuttings such as mums or geraniums, commercial-type seed starters are available with two, three, or four shelves. Thousands of seedlings can be started in one of these compact, easily tended units, which can be used at other times for growing African violets or other flowering or foliage plants. The George W. Park Seed Company Inc., Greenwood, S.C. 29646, lists these units in its wholesale catalog, and has many other worthwhile products for the propagator.

Commercial seed starters can be used for thousands of seedlings and require only a few feet of floor space.

By hanging two-bulb fluorescent units over a table, the home gardener can make his own seed starting unit.

Heating Pads and Cables

Asphalt pads with electric heating elements enclosed in them are better than cables for indoor use. They are convenient to use and, with a proper thermostat, provide a safe, easy way to control heat. Because they are rugged, they should last indefinitely with good care. They are expensive, however, and since those available are not large (approximately 16" by 70"), they can service only a limited number of flats.

Should Have Thermostat

The less expensive soil heating cables work well in hotbeds or in the bottom of a greenhouse bed filled with soil. Because of their tendency to get tangled when you are installing or moving them their life is sometimes short, I've found. Some have a built-in thermostat set a 70 degrees F., which is a great convenience, because if a thermostat is not already built in, you must supply one.

Protect the Cable

Before installing a soil cable in a greenhouse bed, cover the bottom with a layer of several inches of sand or soil. Outdoors in a hotbed, it can be placed directly on the ground. Lay out the cable, following the pattern received with it, and cover it with an inch of sand or soil. Then place over it a piece of screen the size of the bed, to prevent you from digging into the cable when you transplant seedlings. Finally, cover the screen with three or four inches of artificial soil mix, or whatever medium you use for starting the seeds. This top layer must be moistened before seeds are planted.

Propagation Box

Some gardeners use a covered aquarium or a small box covered with plastic as a miniature indoor greenhouse. One or more light bulbs can supply the extra heat needed, depending on the necessary temperature and the size of the unit. The box should be tight to insure the high temperature and humidity that are essential for starting cuttings. Keep a thermometer inside and check it often. A thermostat to turn the lights on and off is a great convenience. Propagation boxes are very useful for starting small numbers of cuttings and for healing evergreen grafts.

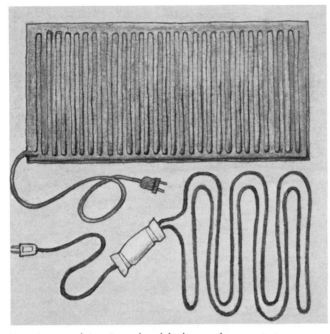

Heating pad (top) and cable have thermostats, so are safe, dependable units for providing heat for seedlings.

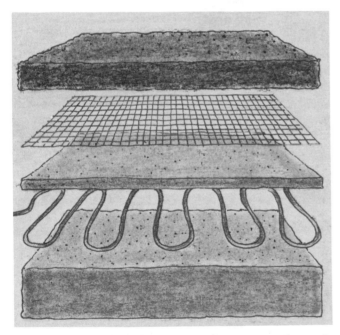

In a hotbed, the layers are, from bottom, sand or soil, cable, an inch of sand or soil, screen, and soil mix. The cable must not cross over itself.

Hotbeds

Several years ago, before I discovered grow lights, I started most seeds in hotbeds. They were crudely made structures, with rough boards nailed together to make rectangular frames the size of our storm sashes. The frame stood about fifteen inches above ground level, and as much below, with the back of the frame eight inches higher than the front, sloping toward the south. I lined the sides with black plastic, and covered the storm window with a sheet of clear plastic for more tightness and warmth.

Although some of our neighbors still used the old method of heating by fermenting a layer of horse manure in the bottom of the bed, I had been told it was difficult to regulate the heat without burning the seedlings, so I installed one of the new-fangled heating cables. Seeds planted in early spring came up quickly, even though the entire frame was covered with several inches of snow.

Choice of Materials

Light-transmitting materials of the type used for greenhouses, such as fiberglass, acrylic, and vinyl sheets, can also be used for covering hotbeds. Many new materials are also available, including rigid plastics made with an insulating air space sandwiched between two layers.

Before the national railroad system made it possible to ship food from the warmer states year round, much of the produce for large eastern cities was grown in hotbeds and cold frames on suburban farms. Some of these plantings consisted of acres covered with glass. Since automatic watering, ventilation, and fertilizing were unknown, a great deal of labor was also involved. Today, with plastic pipe and films, plus modern technology, one person can look after a great many hotbeds, and produce a lot of plants in a small area.

Advantages and Disadvantages

I like many features of hotbeds. On warm spring days you can raise the sash, which helps to produce well-acclimated plants that can later be set outdoors with very little transplant shock. The investment for a hotbed is small, and it takes only a small amount of electricity to operate it. A great many plants can be produced quickly in a few small beds.

Unfortunately there are also disadvantages. The plants sometimes need watering on days when it is too cold or windy to open the frames. Sometimes plants become too large for their space before it is time to transplant them. The worst aspect is that they are demanding. On a changeable spring day, you must be nearby constantly to open the frames whenever the sun comes out, and close them when a cloud appears or it suddenly begins to snow. I began to think seriously about building a small greenhouse after I lost a planting of hybrid tomato plants one day when I was gone for an hour and the weather conditions changed.

Cold Frames

Cold frames are simply hotbeds without artificial heat. They are especially useful for starting plants such as lettuce, the cabbage family, and others that can stand some cold, but need protection from hard frosts. In a cold frame, plants can be started several weeks earlier than outdoors, and it costs nothing for heat or extra light.

I find them useful for starting seeds of perennial flowers and certain trees and shrubs. They are also a good place to harden off transplants or cuttings that have been started in the house or greenhouse. By placing a few narrow boards or slats on the ground in the bed, the pots and flats are kept off the soil. A board base also helps the drainage and keeps the plants clean.

Cold frames are also an excellent place to overwinter semi-hardy plants. Pots and flats of newly started perennials, herbs, and small shrubs

The use of a plastic top on this cold frame makes it easy to ventilate the cold frame and to work in it. A cold frame like this is simple to make.

that cannot stand the direct attack of winter can be placed in a cold frame and covered with evergreen boughs or dry leaves. Although it may freeze inside the bed, the temperature never gets as low as it does outside, nor does it fluctuate as rapidly.

THE POTTING SHED

Although a potting shed is not an absolute necessity for a home nursery, it is a great convenience. It can be as simple or elaborate as you like. In addition to planting and transplanting there, you will find it a good place to graft and do other chores that are messy for the garage or basement. It will be invaluable as a storage space for tools, fertilizer, spraying materials, propagating supplies, records, catalogs, reference books, pots, flats, and much more. In the spring and during bad weather, you will probably spend a lot of time there, so make it comfortable, warm, and well lighted.

If you have a greenhouse, ideally the potting shed should be attached on the north or west end, so it won't shade the house. If possible, you should be able to enter the greenhouse through the potting shed, to avoid having a greenhouse door that lets cold outdoor air flow directly over the plants.

Workbench

The workbench is the most important part of a potting shed. You will probably work there for long periods, so it is very important that the height be right for you so you won't strain your back or neck. Decide in advance whether you want to do most of your work sitting or standing. With the right size stool, one height can be ideal for either position.

The bench must be large enough, but not too wide to reach across comfortably. If possible, build it with edges, about six inches high on three sides to hold any soil that might spill and to

This workbench in a potting shed meets many requirements. It has roomy storage space underneath, several shelves on the back, and a work area enclosed on three sides.

prevent things from falling off the sides or down in back. It should be built of durable wood, and the top covered with a sheet of plastic or Formica counter top, to keep out the moisture. Because high humidity is likely, the nails used in the bench, as well as all the rest of the potting shed, should be galvanized or aluminum.

Build plenty of shelves, bookcases, and bins nearby, so that supplies not currently in use will not clutter the bench. Provide a locked cupboard for pesticides and other dangerous chemicals. A large wastebasket or garbage can is indispensable, and you may want to install hooks on the wall for work clothes.

A window and electric lights should be placed strategically for good lighting on the bench, and plan to install several electrical outlets for things such as a space heater, heat lamp, starting pad, grow lights, and perhaps even a radio and coffee maker. Unless the shed is near the greenhouse, you will also want running water to use for the plants, and a wash-up sink for yourself. Propagation is messy work.

GREENHOUSES

Even a small, simple greenhouse is invaluable for plant propagation, and is especially useful for people who live where growing seasons are so short that they must start their favorite foods and flowers indoors.

Greenhouses come in innumerable varieties and sizes, from expensive pre-fab houses to a reach-in window attached to one's home. My first greenhouse was a modest structure consisting of some cedar poles cut from our woodlot, and covered with polyethylene. An elderly box woodstove provided the heat, and the arrangement worked surprisingly well.

Since then, I have built several other greenhouses, including an A-frame type, one partly underground and part above, a free-standing type, and one attached to a building. They had a variety of roof styles, and were made of many different materials. Each building was a "learning experience." For example, I learned not to make the pitch of the roof too flat in our heavy snow country, and to construct the sides high enough so that when the snow slides off, it has a

place to go. I had trouble with snow accumulation on the A-frame, also, and on the house that was partially underground. I found that in both of the latter, space was limited and there was little headroom.

Placement of the Greenhouse

Because a greenhouse depends so much on the sun for its heat and light, especially at the time of year when the sun is not at its highest point, locate it where it will collect the most light at the times you will be using it. If you operate it all winter, place it where it gets primarily the southern sun. If you intend to use it only in late winter, spring, and possibly fall, angle it so it will get a bit more eastern sunlight.

For use throughout the winter, a house that runs east-northeast to west-southwest will let in more low winter sun. A greenhouse used only in late winter and spring benefits greatly from the rising morning sun which warms it quickly after the cold night. By afternoon the house is thoroughly warm and needs the sunlight less. For the best light in spring, the house should be oriented northeast to southwest, if possible.

Before you build, check the surroundings carefully to be sure the greenhouse is not shaded by buildings or trees, at present, and that no young trees belonging to you or your neighbors will grow up and shade it in the future.

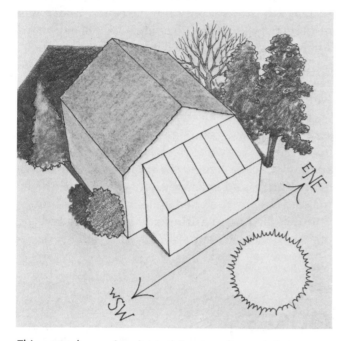

This greenhouse is oriented for use during winter, when sun is low.

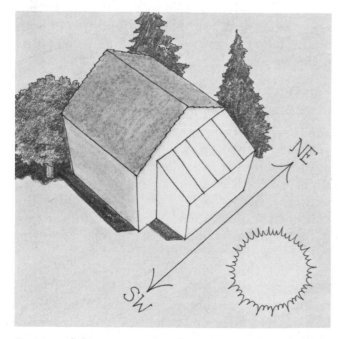

Positioned this way, a greenhouse gets its best light in the spring.

Uses of a Greenhouse

Consider your future needs carefully as you design your greenhouse, because the size of the building and materials used for construction should depend on its use. Will you use it mostly for propagating or for growing large tropical plants as well? Will it be used only in the spring and fall or year-round? Do you plan to use it for things such as ripening tomatoes and pumpkins or drying beans, hops, or seeds in the fall? The off-season uses of propagating greenhouses are endless, we've found. They are a good spot to bake away a mid-winter cold on a sunny day, to dry paint, dehydrate food, wax skis, and even thaw out a frozen Christmas tree. It is such a pleasant place to spend a cool, sunny day that many gardeners wish they had included more room for relaxation.

Swimming Pool

A few years ago, we installed a small, above-ground swimming pool in one of our greenhouses by building the house around it. It helps our propagation effort by supplying lots of humidity for rooting cuttings. It also stores up solar heat during the day, and releases it in the evening, making the heat more evenly available throughout a twenty-four-hour day. The bonus is that we can take a quick dip in late winter, and on chilly days in summer when a trip to the beach is uninviting.

There are disadvantages to having an indoor pool, however. The humidity is so high that everything made of wood rots far more quickly than it otherwise would, and nails and other steel objects rust rapidly. If you decide to install a pool, consider an aluminum frame for your greenhouse.

Greenhouse Foundations

Many prefab greenhouses are available that require no foundation. Some sit directly on the ground, and others are constructed by using pipes that fit into other pipes that have been set in the ground. Some are made to be placed on concrete blocks or treated wooden posts set upright.

If you build a glass house, you must be extremely careful about the foundation. Because glass is so fragile, it cannot stand the shifting

Small greenhouses like this one are ideal for starting seedlings in the spring. Experienced gardeners find other uses for them, such as growing greens in late fall.

caused by frosts on the base of the structure. Consult a local contractor who can advise you about frost depths, and build accordingly.

Although plastic and fiberglass houses are less demanding about foundation requirements, if you are building in a windy area or need year-round protection from cold, an insulated masonry foundation may be necessary.

Greenhouse Coverings

Although polyethylene has increased in price many times since I built my first greenhouse, it is still the cheapest material for covering. Poly comes in several thicknesses—two, four, and six mils are the most common. The four-mil weight is usually used as an outside covering, and the two or four mil, as inside lining. The six-mil poly is stronger and flutters less in the wind, but it is more expensive and no more resistant to the sun and weather than the four-mil.

Gardeners who use their buildings for only a few weeks in the spring like polyethyelene because they can dismantle the structure and get it out of the way for the rest of the year. Although I gave up using it as an outside covering several years ago, I still like it as an insulating liner stapled inside the greenhouse, and for making temporary cold frames and other shelters for hardening off plants that have been started under mist.

Bad Points

I stopped using poly as an exterior covering because it always disintegrated into small pieces by late summer, and blew all over the neighborhood. It was also very noisy on our frequent windy days and, in spite of being well anchored, it frequently tore apart. Spreading out the huge plastic sheets and fastening them down was difficult if there was even a slight breeze. I always planned to re-cover the greenhouse early in the morning on a wind-free day, but soon learned there was no quicker way to start a small hurricane than to start stapling sheets of plastic to the greenhouse rafters.

The development of a type of polyethylene that is more sun-resistant has made it more enticing as a greenhouse cover, and certain kinds now last for several years. Some greenhouses are now being made with two layers of plastic. The space between them is kept filled like a balloon by an air pump that runs continually, and the air space provides insulation and cuts down on the noisy flapping of the plastic. Clear vinyl and other plastic films have also been developed that are more durable and more transparent than poly. These not only let in more light, but also save the annual expense and work of covering the house.

Switched to Fiberglass

As soon as fiberglass became available, I changed the covering on my greenhouse. After handling the flimsy polyethylene, it was a delight to saw and fasten down the large sheets of corrugated fiberglass. The rigidity of the material added to the strength of the house, and I thought I had a covering that would last for years.

After three years, however, I noticed that the geraniums and petunias were not blooming well in the spring because only about half as much light as usual was reaching them. The fiberglass had become a dull, brownish-yellow color instead of frosty clear. I tolerated it for a couple of years more, and then ordered a shipment of high-quality fiberglass guaranteed to last for twenty years. Since it had only recently been developed, I wondered how the manufacturers could be so sure, and my wariness was proven justified. Although it did last longer than the first covering, after about seven years it, too, became badly discolored. In spite of the use of a liquid treatment that was designed to prolong its transparency, it continued to become more yellow every year.

Other Materials

In addition to fiberglass there are numerous acrylics, vinyls, and other new materials that are being used as coverings, but not all are ideal. One acrylic that we put on our greenhouse door shattered into small pieces when we inadvertently let it slam shut one day when it was 20 degrees F. below zero. Some of the thick, very expensive plastic greenhouse glazing panels that our friends have used warped out of shape after a few years of use. Try to find out as much as is known about any new covering and get a meaningful guarantee, if possible, before you reach for your checkbook.

Several new products are being tested that look very promising. Some are very tough, impossible to tear, and appear to be resistant to weather. Because they transmit, rather than absorb, ultra-violet rays, plants are healthier, and you can pick up a winter suntan, as well.

Glass

Glass is the traditional greenhouse covering, and one of the best, because it lets in more light than other materials. It doesn't yellow or warp, and houses made from it always look good. It is fairly expensive, however, especially when the recommended high quality, double-strength, insulated glass is used. It also needs a permanent foundation that doesn't move with frost; and it is vulnerable during wind storms and neighborhood baseball games.

Many gardeners prefer to buy a greenhouse that is ready-made, rather than build it themselves or have it constructed by a local contractor. They are available in many styles and sizes, ranging from small, bay-window types that are attached to the house, to free-standing ones, large enough to grow a commercial crop. Available with them are such conveniences as automatic watering devices, and heat and ventilation controls that make it possible for the greenhouse to run itself, if you must be away.

The appendix lists suppliers, as well as books that give additional information about greenhouse construction and operation.

Greenhouse Interiors

Greenhouse tables or shelves, traditionally called benches, should be constructed of outdoor-type plywood or planks, preferably of cypress or cedar since these are more durable. The high amount of moisture in a greenhouse makes them vulnerable to rot, even if they are painted. A wood preservative helps but it should be a kind that is not harmful to plants. Products such as Cuprinol, which contain copper naphthenate, are considered safe, and are available in both clear and green. Creosote and similar products should not be used as preservatives since they are toxic to plant life.

Benches are sometimes constructed of heavyweight fiberglass. Corrugated sheets are fine for large flats, but the smooth kind is preferable if small pots are to be placed on them. Polyethylene spread over wooden benches is sometimes used, too, but this sometimes bunches up and seals the bottom holes on flats and pots and prevents good drainage.

Hardware cloth made of heavy wire, supported by a wooden frame, is used in some greenhouses very satisfactorily. This type of bench allows excellent drainage and permits loose bits of soil and fertilizer to fall through the wires and not clutter up the bench.

Benches are often built on a slight slant, so water won't stay on them or run off the back while the plants are being watered.

They should be built so they are the right height for working easily, and a width convenient for reaching across. A greenhouse ten feet in width usually has a center aisle three feet wide, with a 3- to 3½-foot-wide bench on each side. A house that is twenty feet wide usually has one central bench, about six feet wide, two aisles and two side benches.

The floor should be of concrete or some other material that is resistant to water, soil, and the chemicals that will frequently fall on it, and it should have a good drain. Greenhouses that have no permanent foundation often have earthen floors covered with several inches of crushed rock. This type of floor allows good drainage, and is especially good in areas where frosts might crack concrete. They are not easy to keep clean, however.

Heating

Greenhouses have traditionally been great consumers of energy. In fact, commercial greenhouses were once built near railroad sidings so carloads of coal could be delivered directly to their sites. In recent years, the price of fuel has prompted a great deal of research into developing better designs and locations of buildings, as well as new methods of heating.

Greenhouses that are attached to a home can sometimes make use of an existing central heating system.

Free-standing greenhouses may be heated in a variety of ways, including hot air, steam, and hot water. A hot-air system is safer than hot water if there is a chance that the pipes might freeze, but it has a drying effect on the plants. Hot water, thermostatically controlled and pumped through pipes beneath the benches, is one of the most satisfactory methods of heating a greenhouse because it produces an even heat.

Some gardeners like to place their heating system in a building adjoining the greenhouse. That arrangement has several advantages. No warm air is lost up the chimney in burning the fuel, no fumes are present to endanger the plants, and the heating plant is not exposed to moisture that would cause rust and corrosion. A steam or hot-water system is most often used, but sometimes hot air can be piped for short distances with little heat loss.

We have heated our greenhouses with a variety of fuels ranging from wood to electricity. Since we have lots of our own wood available, it is the cheapest fuel we have tried, especially if I don't count my labor in the cost. It has never been completely satisfactory, however, and ordinarily I don't recommend it. Wood fires demand a lot of attention, produce an abundance of dust and smoke, and the wood requires considerable storage space. It is difficult to keep wood heat uniform, moreover, and a greenhouse relying exclusively on it is frequently too hot or too cold. Our stove produced an abundance of creosote, even though we were burning dry wood, possibly because the smokestack wasn't high enough. Creosote is bad for plants, and it killed many of our seedlings. Waking up at 2 a.m. to stoke the greenhouse heater was no fun either, and I worried constantly that the building would either freeze or burn down. I knew one old gentleman who slept every night in his wood-heated greenhouse during the spring season, a practical but not very comfortable solution to the problems connected with that method of heating.

Fumes Are Dangerous

A small, tight greenhouse can be heated effectively with a portable electric heater. Kerosene pot burners or gas heaters are also efficient, but they require good ventilation. Unvented heaters should never be used, even with houses that are loosely assembled. They not only consume a great deal of oxygen, but they also produce fumes that are not good for plants or people. Greenhouses covered with plastic or lined with it can be very airtight, which makes them especially dangerous.

Since we use our small greenhouse only in the spring, we try to get by with a minimum of heat, relying heavily on the sun during the day and an oil burner at night. On cloudy days and if the night temperature gets really cold, we place heat lamps over the most tender plants for extra insurance.

A heat sensor in the greenhouse that sounds an alarm in the house is good insurance. It can alert you in case the temperature control system fails and the greenhouse gets dangerously hot or cold.

What Temperature?

Keep in mind your future crops as you plan your heating system. Will you need a cool, warm, or hot greenhouse? A cool house, kept at 45 to 55 degrees F., is fine for starting perennial, tree, and shrub seeds, and is also ideal for grafting potted fruit trees and evergreens. It is the right temperature, too, for starting garden seeds like cabbage, cauliflower, broccoli, and similar plants for later transplanting into the garden. Potted daffodils, tulips, croci, and hyacinths thrive in cool temperatures, as do the leafy Chinese vegetables such as pak choi. A cool greenhouse can also be used in late spring for growing seedlings and transplants started earlier in the house.

A warm house, kept at 60 to 70 degrees F., is suitable for growing most annuals, vegetable plants, and potted plants that are already started.

That temperature is also ideal for starting cuttings from most houseplants, and for rooting hardwood cuttings from trees and shrubs.

A hot greenhouse, 70 to 80 degrees F., is necessary for growing orchids, many cacti, and other tropical and semi-tropical plants.

The temperatures listed are minimum, and on sunny days, the thermometer will climb much higher. Within a tall greenhouse there is a wide temperature variation. Plan to take advantage of the diversity and put the cool weather plants near the floor, saving areas at the top for heat-loving plants.

Heat Efficiency

Until recently greenhouses were a luxury only the rich could afford. Not only were they expensive to build, but heating was so inefficient that even with the low cost of energy, it was expensive to operate one throughout the winter. Even so, there seemed to be little interest in making a greenhouse less wasteful, and often the approved method of removing heavy loads of snow from the roof was to turn the heat high enough to melt it off.

In recent years, as energy costs have skyrocketed, not only have better covering materials been developed, but most free-standing greenhouses are no longer glazed on all sides, since very little heat comes through the north and northwest sides of the house, but a great deal is lost there. Houses are also designed and located to collect as much winter sun and as little north wind as possible.

Heating the Soil

Some growers heat the soil beneath the plants rather than the air, since it costs less. Heating cables, and steam or hot water pipes that run underneath the benches, are among the ways to accomplish soil heating. One grower hooked up a thirty-gallon electric water heater to his greenhouse beds, and added a thermostat and a circulating pump. As the air around his plants cools, the pump comes on, circulates water from the heater, and warms the bed. He says it is very efficient, since it runs only when needed. On cold nights he covers the beds with sheets of plastic to keep the heat contained, which saves even more fuel.

For a most efficient operation, the greenhouse should be kept as full as possible at all times. The more material there is to soak up and hold the heat from the sun, the longer it will stay warm, because heat collected during the day will be released at night. Since we usually have very few flats in our greenhouses when we start to heat them in early spring, we fill the empty spots with potting soil, bales of peat moss, and even garbage cans filled with water. Air alone does not hold heat well.

Fan Is Useful

A blower or fan that keeps the air constantly moving day and night not only provides good air circulation for the plants, but it prevents heat from collecting at the top of the greenhouse while the floor stays cold. On sunny days the blower helps move warm air even to the area under the benches. It warms the soil stored there, which absorbs heat that it releases during the night.

Ventilation

Because greenhouses collect so much sunlight, they sometimes get extremely hot on a bright day. Since plants wilt and die in high temperatures, some provision must be made for ventilation. It can range from opening windows or doors to having an exhaust fan that blows the hot air outdoors or into another building. In spring, because bright sun and heavy clouds are likely to alternate quickly, ventilation by doors and windows demands a great deal of attention. A fan that triggers on automatically when the temperature becomes too warm saves a great deal of labor and worry and takes care of the ventilation problem when you are away.

Since a fan blows a lot of air out of the greenhouse, you must provide a way to let outside air replace it without making a cold draft on the plants. One of the most common air-inlet devices is a plastic sleeve.

For one of our small greenhouses in which the exhaust fan is only a foot in diameter, we cut a sheet of plastic three feet wide and twenty-four feet long, about two-thirds the length of the entire house. This was stapled at the edges to form a long tube about a foot in diameter, with one end closed. The other, open end was secured

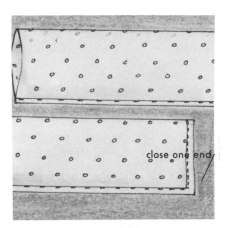

For a draft-free air inlet, make this tube out of a sheet of plastic. Cut dozens of small holes in it.

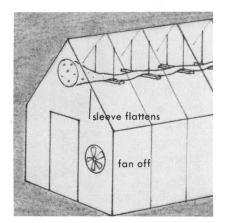

Tube should be hung at the top of the greenhouse. The open end is secured to an opening in house wall.

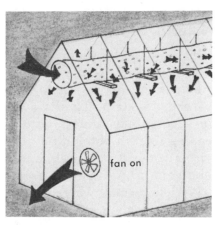

Then exhaust fan is turned on, fresh air pulled in through tube spreads quickly into greenhouse.

to a round opening of the same size, located near the peak of the greenhouse. All along the tube I punched dozens of small holes, about a half inch in diameter.

The tube is supported by boards hung from the peak of the house. When the fan is not running, the tube hangs limp across the boards, shutting out any cold air that might otherwise enter the opening. When the exhaust fan triggers on, however, the sleeve automatically billows full and lets in air from the outdoors. Because it enters along the top of the house, no drafts or air currents can hurt the plants. On very warm days in late spring, we shut the fan off and leave the doors open, but the fan nicely takes care of ventilation on the changeable early spring days.

Shading

A greenhouse is shaded both to reduce light for certain plants and to lower the temperature. It can be as challenging to keep the building cool in summer as it is to keep it warm in winter. Painting the greenhouse at the beginning of the summer with a white paint made especially for shading greenhouses was considered the best method for years. The paint cannot be removed from fiberglass and other plastics as easily as from glass, however. Plastic shadings of various densities, available at nursery supply houses, are now being used for shading. They can be quickly installed and in a manner that allows them to be partially removed on cloudy days when more light is beneficial.

Water and Electricity

Water and electricity are necessities, rather than conveniences, in a greenhouse. In addition to using electricity for lights, ventilation, and automatic controls for your heating system, you may want to use it also for bottom heating of hard-to-root plants, emergency heating, an alarm system in case the temperature control system fails, and possibly a radio for your own pleasure and that of your growing plants.

I bury the electric wires leading to our free-standing greenhouse, since I feel that they are dangerous overhead in an area where we are likely to carry around ladders and metal pipe.

Bury the Water Line

If your greenhouse is not attached to your home, be sure that the water line is buried deep enough so it won't be frozen when you need it most, as you start plants in early spring. In my first greenhouse I failed to do that, and I spent hours carrying water to the plants for a month, while waiting for the buried pipe to thaw out.

When I remodel our greenhouses, I will include enough faucets so I can reach all the plants with short hoses. A long hose is a nuisance because it always gets tangled and is underfoot. Many greenhouse operators hang their hoses on hooks that slide along an overhead pipe in the manner of a shower curtain. That way the hose won't fall on the floor and pick up disease that may later be sprayed over the plants.

Watering Automatically

Plants need frequent waterings, especially in sunny weather, so if you must be gone during the day, you may want to install an automatic watering system. This is available from most nursery supply houses, and has a small tube leading to each pot and flat. It can be turned on with a timer if you wish, and it will effectively tend your crop while you are away.

We use a small watering can and a sprinkler bulb rather than a hose when watering seedlings. The force of the water from a hose is too strong for delicate plants, and their fragile stems are easily injured, or their roots washed out. When the plants are larger and stronger, we use a hose with a nozzle that breaks the water into a gentle mist.

Solar Water Heater

Because icy water is not good to use on plants, it is necessary to have some method of warming it. We use solar heating, and keep a 200-foot coil of black plastic pipe in the peak of our greenhouse, which absorbs the sunlight. Because the water often gets very hot, I installed a valve so it could be mixed with cold water. The heating coil holds enough for one watering, and by the time

If you're forgetful, or if you're away from home a great deal, consider a watering system like this. It's automatic, controlled by a clock, and provides water to each plant through a system of small plastic tubes.

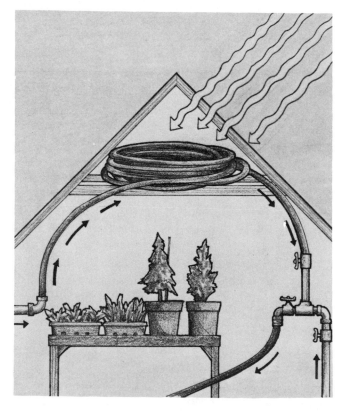

When seedlings have grown for a few weeks, they can be watered using a nozzle like this that breaks flow of water into a light mist that won't damage the plants.

On a sunny day, water in this simple solar water heater will get very hot, so a mixing valve must be provided.

another is needed, it is warm again. Even on days when the sun does not shine, enough light is absorbed by the dark coils to warm the water sufficiently.

Sanitation

Plants should be closely inspected frequently for possible insect and disease problems. The warm, humid atmosphere in a tight greenhouse makes an ideal breeding place for all kinds of troubles. Insecticides and fungicides should be kept handy in case you need them. Always use them carefully, following the directions on the package.

In a tight greenhouse, spraying and dusting must be done with far more care than would be necessary outdoors. Leave the area immediately after spraying and don't reenter until the air has cleared. Make sure the family dog or cat isn't curled up under a bench taking a nap before you begin, too. Our cat feels the greenhouse is her private property.

CONTAINER GROWING

Most varieties of plants grow well in containers, if the proper growing conditions are provided. Growing plants in pots outdoors instead of in nursery rows has many advantages. It is useful if you don't have room enough in the ground for all you would like to grow, since many more can be grown in the same area in pots than in nursery rows. Container growing is also worthwhile if you would like to sell a few plants but don't want to be bothered to dig every time someone wants one. You can buy soil and mix it for the pots, and not lose it from your nursery area, as you do when plants are dug and wrapped with a rootball. A further advantage is that a potted plant is easy to sell and convenient for the customer to take home and plant. Its roots are intact and it is almost certain to live.

Container growing does have its disadvantages, however. The plants require daily watering whenever it doesn't rain hard, and additional fertilizing because the increased water leaches nutrients and lime from the pots. The wind sometimes topples them over, and certain varieties may not overwinter well on top of the ground in areas where temperatures are low, unless the plants are buried in the snow.

Plants Need Full Sun

Although nursery slatted houses look nice, in most cases potted plants should be grown in full sun. They should be placed on coarse gravel or crushed rock for good drainage, and checked frequently for disease and insect trouble.

The soil used in the containers should be por-

ous enough for good root development, but still hold water and fertilizer well. The usual mixture is (by volume) one-third loamy soil, one-third sand, and one-third peat moss or composted bark. The proper amounts of fertilizer and lime should be added and all should be thoroughly mixed.

Different plants have different requirements, but a vague rule is, one cup of lime, and one quart of dried manure to each wheelbarrow load (about four cubic feet) of soil. Slow-acting fertilizers such as Mag-Amp can be used instead of the

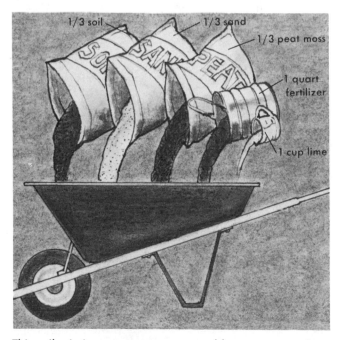

This soil mix is easy to prepare and has proven to be excellent for a broad variety of container plants.

manure. Follow the directions on the bag concerning the amount. A wetting agent such as Triton may also be useful in the soil of container plants to help absorb water better, instead of letting it run through the pot. Covering the top of the soil with a mulch of shredded bark helps prevent the soil from being washed away by rains or watering.

Watering can be done with a hose or sprinkler can, and the plants should be checked frequently to see that they are not drying out. Tip a plant over occasionally, and tap it from the pot to make sure that the water is soaking to the bottom of the roots.

Weed Problems

Weeds are usually no more difficult to control than when the plants are grown in nursery rows. Sterilized soil reduces the number of weeds, and mulches or fiberglass mats can be used over the soil in pots to prevent seeds that have blown in from elsewhere from sprouting. Herbicides, such as Dacthyl and Treflan, may also be sprayed on the soil to suppress weed growth.

Since watering leaches away the fertilizer rapidly, extra amounts of plant food should be added during the growing season. Use manure dissolved in water, or a liquid chemical fertilizer such as Rapid-Gro every two weeks until the middle of the summer. Dry chemical fertilizers, such as 5-10-10, can also be used, but like all chemicals they must be used carefully, since excessive amounts will cause fertilizer "burn."

Hold a plant gently in one hand and tap it from the pot to see whether water is reaching root bottoms.

Plants usually grow much faster in containers than in the ground, because the feeding and watering can be better regulated. A lilac that might require three years to grow to saleable size in a nursery row grows to the same size in one year in a large container.

Do not allow plants to remain in the pots long enough to become rootbound. When transplanted, the roots will continue to grow in a circle and strangle the plant.

CHEMICALS

No matter how one feels about the use of chemicals on plants, most serious gardners find that sooner or later they must resort to them. If we sell plants, the law requires that they be free from disease and insects, and even if we don't sell, we want healthy plants.

Pesticides come in many forms, and the most common are insecticides, fungicides, and herbicides. They range from a simple dusting of sulfur to control blight, to powerful systemic controls that kill certain insects. This list includes such things as mouse poison, deer and rabbit repellents, and growth regulators.

Probably, for a small operation, you will be able to buy the few chemicals you need quite easily. You may find, however, that some of those recommended by your Extension Service for a special problem are regulated, which means that you cannot obtain them without a license. Ask your extension agent about procuring such a license. Most states require that you enroll in a short course and take a test, to prove you can safely handle chemicals.

Insecticides have been used since man found it was possible to control certain insects by sloshing soapsuds over his garden occasionally. They range from such powerful chemicals as DDT and Parathion, now carefully regulated by law, to the mild pyrethrums, thuricides, and soaps that are favored by organic gardeners. Some insecticides help control a wide range of insects, while others have been developed especially to annihilate only a few specific pests. Some most used by gardeners are carbaryl (Sevin), Diazinon, Imadin, Malathion, rotenone, and methoxychlor. Thuricide is preferred by organic gardeners, because it controls many chewing larvae by destroying them biologically rather than chemically. Recently, systemic insecticides have been developed which are absorbed by the plant. Cygon, Temik, and the Disystons attack the bugs as they eat the plant. Their use is quite restricted, especially for food plants. Although they are effective, they must be used very carefully, especially in a greenhouse.

Choose Carefully

The kind of insecticide you choose to use will probably depend on your experience, your ideology, and the recommendation you receive. Insecticides vary greatly in toxicity, but no pest killer, even those endorsed by organic gardeners, should be considered safe. Follow the directions on the package carefully.

The fungicides that attack diseases are much newer chemicals. Lime sulfurs were among the first used, but copper sulfate and others soon followed. These must be used carefully or some damage to the plant will result. Consequently, the original fungicides are less commonly used today than the newer developments such as benomyl (Ben-Late), Captan, Dexon, Ferbam, Subdue, and Zineb. One or another of these products is found in most common garden dusts.

Fumigants such as chloropicrin, methyl bromide, and Vapam are used for sterilizing soil and seed beds. These sterilants can be quite dangerous, however, and should never be used without a thorough understanding of the process.

Growth Regulators

Growth regulators are chemicals used to stimulate plant growth or to stop plants from growing temporarily. Gibberellic acid, for example, is used to stimulate the growth of grapes, citrus, and other crops, and to cause mutations in seedlings. B-Nine SP and similar chemicals are used on potted and bedding plants as growth retardants to keep the plants from getting tall and leggy. Petunias, poinsettias, chrysanthemums, and others are often treated with these.

Herbicides are weed killers, and there are so many of them now that they have replaced insecticides as the most used pesticides. Like the insecticides, many were developed for a single purpose, such as controlling annual weeds in field corn or getting rid of grass among woody plants. If you are thinking of buying one, check the label, and buy it only if it is meant for the job you want it to do. If you are considering the use of herbicides, ask your Extension Service for the weed control chart for nursery crops. It will give you much valuable information, and include the latest information on each chemical.

One of the first herbicides was 2-4D, the miracle chemical that made possible a weed-free lawn. This and many of the other weed and brush killers work through the foliage and must be sprayed on the plant when it is actively growing. Some foliar weed killers are selective and can be sprayed over nursery stock to kill certain weeds and grasses without hurting the good plants. Others must be used as a directed spray because they kill a wide range of both wanted and unwanted plants.

Other Herbicides

Some of the other herbicides that work through the foliage are ammonium sulfamate, Amitrole, Paraquat, and Roundup.

Herbicides that work through the roots, used mostly to control weeds among established plants, are Atrazine, Casoron, Karmex, Simazine, and Sinbar. Like the foliar sprays, these must be carefully used, since they are effective for only certain weeds, and are sometimes safe to use only on a limited number of plant varieties.

Another group of herbicides works through the soil to restrict the growth of sprouting seeds. They can be used on plants that are already well established and growing, either in the ground or in pots. They help prevent the growth of weed seeds already in the soil, or any that blow in. When used correctly, these are great labor-saving chemicals, but they vary considerably in the length of time they are effective. Some last only a few weeks, while others are effective for most of the season. Dacthyl, Devrinol, and Treflan are a few of these pre-emergence herbicides.

COLLECTING NATIVE PLANTS

A good way to increase your plant collection is to gather some varieties of the excellent plants that grow native in every area. Each region has many worthwhile indigenous plants as well as a lot that have escaped from gardens and set up living quarters in the wild. Seedlings or small plants can be dug and transplanted into nursery beds and later sold or planted out as windbreaks, hedges, ornamental shrubs, shade trees, or to attract birds and wildlife.

If you own pastures and woodland, you have great opportunities for collecting. Scavenge on private or public lands only with permission from the owners, or people in charge, and tell them exactly what you intend to do. It becomes more apparent each year that we should practice wise environmental conservation wherever we go, including on our own properties. Be especially careful not to dig up endangered specimens.

There are many advantages to working with native plant life. You are sure that the plants are hardy and that they will require less care than exotic species. They are fresher, too, than plants you might buy by mail, many of which were dug the previous autumn, stored all winter, and sent to you in the spring.

Need Right Conditions

We never dig many of the plants we admire most because we don't have the proper environment in which to plant them. Many wild plants require specialized growing conditions, and it would be wrong to gather water lilies, royal ferns, bog cranberries, wildflowers, or native rhododendron if you don't have a spot where they'll survive and thrive. The best rule is to observe the soil, moisture, and light conditions where you find each plant growing, and plant it in an environment as similar as possible.

Many springs ago my 4-H club members pulled up wild evergreen seedlings about eight inches tall to sell for reforestation and Christmas tree plantings. If they found a spot where the seedlings were thick and the soil was moist enough to get the roots easily, a half dozen boys could easily pull 10,000 seedlings each day. I have also collected white birch, maple, and beech seedlings, as well as small shrubs and flowering

plants, but these had to be dug with a shovel rather than pulled.

Dig in Spring

Although I always dig plants in the spring, each summer I scout around, searching for candidates for later transplanting. Birds drop seeds from the berries they have eaten in other areas, so each year there are surprises in our woods and pastures, and along the country roads.

Early spring is the best time to dig and transplant wild plants because there is less shock if they are dormant when you move them. Even with the best care, it is difficult to dig wild plants without damaging the roots badly, especially when you are moving larger trees and shrubs.

If I plan to move a maple, spruce, or other wild tree with a trunk larger than an inch in diameter, I always root-prune it in the spring one year before I intend to dig it. With a sharp spade, I cut straight down into the soil, in a circle around the tree. The cut should be a foot or more deep and two or more feet from the trunk of the tree, depending on the tree's size. I dig as if I were planning to move the tree, but without actually lifting it from the hole. Although this procedure removes a lot of small roots, since I don't cut off the bottom roots and the plant is not actually moved, it stands the shock of losing them very well.

New Roots Develop

By the following spring, a heavy new root system has developed, mostly within the cut circle. I then dig the plant by cutting out the same circle, but this time, I slide the spade under the main trunk of the tree as I dig, and cut off the bottom roots. I lift the tree from the earth, wrap the roots well in a sheet of burlap, and take it home. I plant it in the spot I want it to grow permanently and about two inches deeper than it was growing originally.

A tree or plant dug from the wild needs even more loving care after planting than a nursery-grown transplant. Unless the soil stays moist, water it every day for a least two weeks, and add liquid fertilizer to the water once a week. Keep watering once or twice a week for at least an additional month.

A year before moving a small tree, root-prune it with a sharp spade.

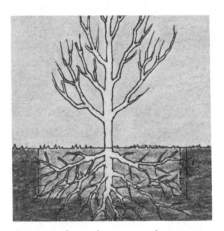

Cut a circle at least two feet away from the trunk and one foot deep.

The next spring, cut again along the lines of the first circle.

At the same time, cut the roots that are underneath the tree.

Lift the tree from the hole, set it on a sheet of burlap and tie it.

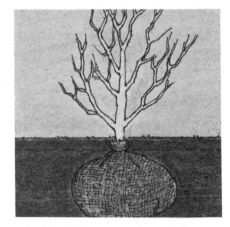

Open burlap, then place tree two inches deeper than it was before.

Digging an extra large wild tree is likely to be rather unsuccessful, unless you have special equipment. Keep in mind that there is nearly always as much of a tree underground as there is above. Larger trees or shrubs require moving a lot of soil, and a proper rootball could weigh several hundred pounds. A three-foot to five-foot evergreen or nut tree, an eight-foot to ten-foot deciduous tree, and a three-foot to four-foot shrub are the largest sizes most amateur gardeners should attempt to transplant. Small trees and shrubs can be moved with less shock to the tree, and they begin to grow sooner than large ones that have suffered extensive root damage. Often a young tree, carefully moved, will quickly overtake a larger one in growth, and within a few years develop into a stronger and better specimen.

SEEDS

Every time you plant a seed you are carrying on an ancient tradition. The earliest written records refer to the sowing of seeds and the harvesting of crops. It is easy to visualize an imaginative primitive creature puzzling over the sprouts coming from seeds in the garbage heap outside his cave. Probably he wondered if it might be possible to help nature along, and grow his food nearer home.

Although over the centuries man has discovered many other methods of propagation, planting seeds is still the most common. Grains, grass, vegetables, weeds, nuts, forest trees, and most flowers and herbs are grown from seeds, as are many perennials, shrubs, and grafting stock for fruit trees.

Inside a Seed

A seed is an amazing little package. Each contains within its coating an embryo plant and a food supply. Given the proper conditions of moisture, light, oxygen, and temperature, the seed will lose its dormancy and grow.

Each kind of seed demands specific conditions for germination. Certain types, like black walnuts, must be frozen before they sprout easily. Some cactus seed, on the other hand, need a warm, dry atmosphere to sprout. The seeds of a water lily should be completely submerged.

Most trees, shrubs, vegetables, and flowering plants have two separate seed leaves and are called dicotyledons. Others, such as grasses, grains, palms, lilies, pineapple, and asparagus, that have only one leaf, are called monocotyledons. The difference is clearly visible as the seed sprouts. A root emerges from a dicotyledon seed, the sides of the seed divide, and inside are two leaves. This process is plainly illustrated in the way a bean sprouts. Monocotyledon seeds, like corn, send one sprout upward to become the top, and another downward to grow into roots.

Seeds develop as a result of the pollination of the female part of a flower by the male pollen of the same or another flower. That process, and how it can be used in developing new varieties, will be discussed later in this chapter.

Monocotyledonous plant sends one sprout upward to become the top, and another downward for roots.

Dicotyledonous plant sends down root sprout, then seed divides to permit two leaves inside to expand and grow.

ANNUALS AND PERENNIALS

New gardeners sometimes have trouble with the traditional terms used for describing garden plants. An annual is a plant, such as a marigold or radish, that is planted in the spring, grows one season, produces flowers and seed, and then dies. A biennial, like sweet william, or black-eyed susan, grows the first year from seed, blooms the second, and then dies. A perennial, such as a geranium or peony, lives for a long time, often for many years, although it may not be able to survive a winter outdoors in a cold climate.

This terminology can be confusing because some plants that are considered biennials in one area may live for several years in another, and plants that are tender perennials, such as petunias, are often grown as annuals. Trees, shrubs, and woody vines are perennials, but the term is used mostly to refer to herbaceous plants that die to the ground each winter, so their stems never become woody.

To be useful to a farmer or gardener, a seed must come "true." Anyone who plants a lot of seeds wants to know exactly what kind of crop that seed is going to produce. Seeds of the crops most commonly planted have been standardized for years, and years of inbreeding insure that every beet, pea, or marigold seed in a package will produce plants more or less alike. Native grass, forest trees, and most wild flowers and shrubs have standardized themselves, and their seeds produce plants very much like their parents.

Many of the plants that have been developed with man's help, however, cannot be reproduced reliably by seeds. If you plant the seed of a Mandarin orange, a Rome apple, or a Peace rose, the seedlings will most likely grow into plants that are quite different from their famous parents.

This fact has often puzzled gardeners. When I was growing up, a superstition I often heard was that one of the ten seeds inside each apple would grow into a tree exactly like the parent. Probably a more accurate statement would be that of the many thousands of seeds produced by all the apples on a tree, only one might produce a tree that closely resembled the parent. Because each "hybrid" seed contains a different combination of the many genes of its numerous ancestors, a resulting tree could resemble any of them, including a wild one, or none of them. A rare seed, however, occasionally produces a tree or plant that is far superior to others in its class, which is how new varieties are born.

Improve by Selection

Just as the different breeds of farm animals have been developed over the years, the quality of various plants has also been improved by natural selection and hybridization. Primitive agriculturalists learned at an early date that it was worthwhile to save seed from the best plants that they grew. They also learned to plant the seed of some of their good plants next to seeds of superior strains they discovered in other areas, and the resulting plants were even better.

The American Indians, for example, by working with a small grain that grew barely taller than wheat, developed corn (maize). By natural selection of seeds and a simple form of hybridizing, they produced not only a vegetable much like our garden sweet corn, but also a field corn suitable for pounding into meal, and a special treat—popcorn. The many different strains of corn developed by the Indians, and later by the early settlers, became standardized over the years by inbreeding, and by carefully not planting the different strains near each other.

WARNING: PATENTED PLANTS

Certain trees and flower plants are patented by the U.S. government, which means that the developer has applied for, and received, exclusive rights of propagation. Asexual propagation of these plants without permission is forbidden by law. Many tea and floribunda roses are patented, as well as certain fruit and shade trees.

Some large nurseries that own patents sometimes sell their plants wholesale to smaller nurseries or garden centers, and occasionally sell others a license to propagate them. When a tree or shrub is patented, the catalog description or label will so indicate. Stark Brothers Nursery has patented a large number of fruit trees, and Jackson and Perkins sells many patented roses.

Although you are not allowed to reproduce the plants by grafts, cuttings, layers, or similar means until the patent expires, the law does not prevent you from planting their seeds if you want to use one of these superior plants in a breeding project.

BUYING SEEDS

Although it is an interesting hobby to save the seeds from our own plants or gather them elsewhere (see "Collecting Seeds"), most of us buy them instead.

Fortunately, there are many good seed suppliers, and if you have gardened for a few years, you probably already receive several garden catalogs. If not, request a few from the companies listed in the appendix. The companies listed issue wholesale lists, so if you plan to grow large numbers of seedlings, ask for a copy.

There is a wide variation in the price of the same kind of seed from different sources, so you may want to try some lower-priced seeds occasionally, to see how they compare in quality with the more expensive ones. The difference is often only in the packaging and advertising.

Some gardeners, feeling they will be planting seeds better acclimated to their climate, choose to buy them from a company nearby. Actually, with only a few exceptions, most seeds sold in this country are produced on the West Coast, or in Europe or Japan, so unless you are assured by a local supplier that the seeds are grown nearby, it doesn't matter where they were packed and labeled.

It is important, however, to grow varieties that are suitable for your region. Seed catalogs don't always supply that information, so check with your state Extension Service or other gardeners in your area if you are in doubt about the best kinds for you.

Buy Early

We order our seeds for spring in early winter, as soon as the catalogs arrive, so we will be sure to get the varieties we want. For the same reason, I pick up any seeds I buy from the local store as soon as they come in. It doesn't help the seeds to be exposed in a warm store for months, so I like to get them home and stash them away in a cool, dry place.

It is easy, in the cold, barren winter, to succumb to the temptation of colorful catalogs and order far more seed than you need. Unless you have a great deal of space and time, and a good use for surplus plants, restrain yourself and keep your operation under control. It is always discouraging to dump out perfectly good plants, simply because you don't need them. Giving them to a neighbor is far more satisfying.

Traveler's Seeds

Visitors to foreign countries often want to bring home seeds they collect or buy there. Usually the Customs Service does not object if you bring in clean seeds that have been packaged commercially, but it is a good idea to check with them first. It is illegal, however, to import uncleaned seed, uncertified bulbs, plants, or soil. There is a strong possibility that insect eggs or disease may enter on them and, as most gardeners know, we have enough pest problems at home already.

GROWING BEDDING PLANTS FROM SEED

Annual vegetable and flower seedlings that are started early in the season, for later transplanting to the garden, are classified as "bedding plants." If you intend to sell nursery stock, this type of operation makes it possible to earn an income from plants you grow and sell the same season.

There are many good reasons for growing your own bedding plants, even if you don't intend to sell them. If you use many plants, the financial savings are worthwhile. You will no doubt be able to grow plants that are larger and healthier than those you could buy, and you have better control over their size when you set them out. If you want plants that are new or unusual, you probably will need to grow them yourself, too, since most sales outlets are rather limited in the number of varieties they carry. Most gardeners admit that one reason they grow their own bedding plants is for the pleasure of working with the tiny green seedlings while winter still rages outdoors.

The vegetable and flower seeds my family started in the house each spring when I was a child always sprouted and grew well on our sunny window sills, and I don't remember that we ever lost any because of disease. When I later became more seriously interested in gardening, however, the seeds often sprouted well, began to grow, and then fell over dead. I had no idea of how to cope with the problem.

Bit by bit, I discovered what I'd been doing wrong. My family had planted tomatoes, cabbage, and marigolds in the rich, virgin soil we had brought from our maple woods each fall. They planted them in our home where three roaring woodstoves kept the temperature above 75 degrees F. day and night.

By contrast, due to high fuel costs for our electric and oil stoves, we had begun to keep our home quite cool, especially at night, and although my seedlings usually got plenty of heat in the sunny bay window during the day, the rest of the time it was far too cold for them. Instead of using healthy soil from the woods, I planted seeds in a garden soil-sand mixture which was loaded with viruses. Furthermore, I was attempting to grow eggplants, celery, petunias, begonias, and similar plants which were much more susceptible to disease than the easy-to-grow seedlings of tomatoes and cabbage.

To add to the problems, the water I thoughtlessly used on the plants was taken directly from our cold water faucet and was only a few degrees above freezing. It kept the soil wet and cold for hours. I also faithfully watered the plants early in the morning, according to all the directions, "because the sun would dry out the soil during the day." Unfortunately, often the weather turned cloudy by 10 a.m., and it rained or snowed, which meant that the plants stayed cool and dripping wet all day. Naturally they didn't grow on those days, and were ripe for attack from the various root diseases that thrive in such conditions.

Commercial greenhouses do a big business in bedding plants every year because so many gardeners have difficulties in starting seeds inside, just as I did. Fortunately the problems of home growing can be easily remedied when the plants' requirements are understood.

Germinating Temperatures

The seeds of most bedding plants germinate best when the temperature is kept between 75 and 80 degrees F. from the time they are planted until the plants develop their first set of true leaves. Try to keep the temperature relatively constant throughout the day and night. If you grow the seedlings near cold windows, move them when it becomes dark outside, so they will be close to a radiator or stove during the night.

After the plants have developed, they grow best in a cooler temperature—approximately 65 to 70 degrees F. by day, and 55 to 60 degrees at night. Some exceptions are the heat-loving plants such as tomatoes, peppers, eggplants, and the concubits—squash, cucumbers, and such, which prefer 70 to 80 degrees F. by day, and 60 to 70 degrees F. at night.

SELLING PLANTS

If your propagation has been successful, you may want to sell some of the results. The selling of plant material is closely regulated, and you must comply with certain state and federal regulations.

Each nursery must be registered, inspected, and certified each year by the Department of Agriculture in your state, which ensures that the plants offered for sale will not harbor dangerous insects or disease. After your operation has been certified, the nursery inspector will notify you if there are other necessary inspections of plants that are currently quarantined in your area.

If you sell by mail, or send plants out of state, a copy of your inspection certificate must accompany each order. You will probably need to register with the state tax department and collect a sales tax, and also register the name of your business with your secretary of state as well as record it locally in either your town or county municipal office. You may also need a zoning permit, and special permission if you erect a sign.

Although these procedures seem rather complicated, they can usually be accomplished with very little time and expense, and without a lawyer. Unless you intend to sell plants regularly for several years, however, it might be easier to give them away and dispense with the bother of starting a business.

Whatever the temperature, keep the plants out of drafts caused by open doors, windows, or fans. The chill factor works on plants as well as people, and they are adversely affected by cool breezes or sudden temperature fluctuations.

Although seeds will sprout in the dark if the temperature is right, they develop well only if they get plenty of light. Imitate the long days of spring and give them as much light as possible for the first two weeks.

The proper amount of water is critical for seedling growth. Too much or too little at the wrong time, or the wrong temperature, can kill them or create conditions for disease, as my bad experience showed. It is a painstaking job to sprinkle a flat of seedlings with just the right amount of water, but the results are worth the effort.

For many years I tried to fulfill all the fussy requirements of bedding plants as I started them in sunny windows, hotbeds, and greenhouses. I used heating pads, soil cables, and plastic tents, all with varying degrees of success. The big breakthrough came when I finally discovered grow lights. Grow lights make it easier for a home gardener, with a minimum of space, to satisfy all the requirements of small seedlings. We have consistently had good luck using them, and they are now our favorite method of starting bedding plants.

If you have not had success in starting plants in your home, you may want to try grow lights. If you prefer hotbeds, windows, or cold frames, you can still adapt the following cultural directions to your method.

OUR GROW-LIGHT METHOD

(1) Pack the soil mix into a sterilized seed flat, making it level and slightly lower than the top edge of the container. We no longer use soil, even sterilized soil from the garden or virgin soil from the maple woods but, instead, plant seeds in a commercially prepared, artificial soil mix. (See "Media.") Because it is sterile, the possiblility of disease is lessened, and there are no weed seeds to sprout and crowd the young plants. Artificial soil mixes hold moisture well, and most contain enough fertilizer to feed the plants for a short time as they are getting started.

(2) Soak the flat thoroughly by sprinkling it several times with slightly warm water. Some mixes contain a wetting agent that helps the mix absorb the initial soaking faster, but even so, most are very dry, and tend to repel water when it is first applied. When we have a lot of flats to soak, I set them in a shallow tray made of plywood that has been lined with plastic and filled with water. Several trays at a time can be set in this soaking box. The mix absorbs water from the bottom while I am doing other things, and in a half-hour or so they are moist enough.

When starting seeds, fill flats nearly full, then moisten by sprinkling with warm water.

Sow seeds in furrows an inch apart. Sow thinly so seedlings won't have to be thinned later.

Cover with perlite, using a thin layer for small seeds, more for larger seeds. Place under lights.

(3) Plant the seeds in shallow furrows an inch apart. Seedlings in rows are easier to dig out and transplant, and there is more air circulation and growing space between them. I always fight the temptation to plant seeds too thick, but, even so, when the seedlings sprout, I wish I had been more stingy. Some companies sell pelletized seeds, which are small seeds enclosed in a claylike pill. These are much easier to space at the proper distance, and some growers prefer to use them. Seedlings grown too close together have weak stems and transplant poorly.

(4) Cover the seeds with a fine grade of perlite. It dries out rapidly, and seedlings do better if the top of the media is fairly dry most of the time. Use a thin layer, so seeds won't need to struggle to get through it. Barely cover fine seeds, like begonias and celery, but put a slightly thicker layer over larger seeds such as asparagus and dahlias. Some growers prefer not to cover the extremely fine seeds at all, but lay a sheet of clear plastic over them, removing it as soon as the seeds sprout.

(5) Label everything with plastic markers, writing with a heavy lead pencil or a marker with waterproof ink, and place the trays under the grow lights.

Lighting

The lights should be about six to eight inches above the tops of the flats at first. Then, as the seedlings grow, they should be raised to maintain that distance. Because the maximum temperature in our home in winter is about 65 degrees F., we place a tent of clear plastic over the whole unit, to contain the modest amount of heat given off by the fluorescent grow lights, and to cut down on drafts.

Leave on the lights for twenty-four hours a day at first, unless the unit is placed where the sun will shine on the flats for part of the day. Although most directions call for less light, I've found that the extra light and heat, combined with careful watering with slightly warm water, encourage fast germination. Some seedlings appear in less than two days, and nearly all sprout in less than a week. The fast germination and speedy growth make them less likely to contract a disease.

After the plants are up and growing well, shorten the light period gradually so that in about three weeks the plants receive only twelve hours of fluorescent lighting each day. I leave them on at night so the seedlings actually get more than twelve hours of light each twenty-four hour period, counting the daylight hours. An automatic clock timer is a big help in regulating the lights.

In about four weeks, move the seedlings away from the lights to a cooler place. We put ours on a table in front of a large window, because they still require at least twelve hours of light each day. We then start another batch of seeds under the grow lights. As soon as the weather is warm enough and we begin to operate the greenhouse, we move them there and begin transplanting. In the days before I built the greenhouse, I set the seedling trays into a cold frame, and covered them carefully on cold nights. Sometimes I had to put in a light on a long cord to keep them warm if the temperature went too low.

Watering

The plants should be examined at least once a day, and more often if possible. Not only will you want to be in on the sprouting of each batch of seedlings, but you should examine them also to see that they are not too wet, too dry, or looking unhealthy. Watch for insects, such as white flies and plant lice (aphids), particularly if your home includes some plants from commercial greenhouses. Too many growers are not extra-careful about insects and pass them around generously. Use a mild garden aerosol spray, such as House and Garden Raid, if these bugs appear.

From the time the seeds are planted, watering is a tricky chore. We use a small plastic, bulb-type sprinkler, similar to the kind used for sprinkling laundry, since the force of a watering can or hose could easily wash out the seeds or break the tender stems. I can't repeat often enough that the water should be barely warm, and never hot nor cold.

Frequent light waterings are better for seedlings than a heavy soaking every day. The top covering of perlite and the seedling leaves should have a chance to dry out between waterings. This procedure is quite different from that recommended for watering older, established plants.

After the lights are on only part-time, I apply the water just as the lights come on each day. The plants are making their fastest growth at that time, and the tops dry quickly under the

Use a bulb-type sprinkler and warm water to keep seed beds moist.

lights. This is preferable to having them wet as they enter a cool, dark period. After the plants are moved away from the lights, watering should be done early in the day, just as the sun (hopefully!) begins to hit them.

Bottom watering was once a common practice. Gardeners took each flat and set it in a pan of water each day. This way they could water the roots but keep the tops of the plants dry. Unfortunately, in order to supply enough water to reach the shallow roots, the soil at the bottom of the flat had to be soggy wet, and this wasn't good, especially since the peat in artificial soils can hold a lot of moisture. Waterings from the top, I have found, give better results with artificial soil mixes.

Fertilizing

By the time the seedlings have been up a week or two, they will need extra nourishment. Although the food stored in the seed and the fertilizer in the mix are enough to start the seedling on its way and keep the plant growing for a couple of weeks, additional fertilizer helps ensure good health and fast growth. Give them a weak solution of a liquid fertilizer such as liquid seaweed or Rapid-Gro and continue to do so about once every week or two.

One year I got the bright idea of sprinkling a small amount of dried poultry manure over the mix to help feed the plants after they got started. Although the manure was nearly odorless in the bag, when watered it created a new atmosphere in our home. We discontinued the practice with the next batch of seedlings we started in the house, but still consider it very worthwhile for plants we start in the greenhouse. The earthy smell disappears after a few days.

Controlling Disease

The biggest challenge faced by anyone starting annual flower and vegetable plants indoors is control of the various soil diseases, described by the odd expression, "damping off." Outdoors, the environment offers some natural controls, but an artificial climate, with high humidity and poor air circulation, invites disease.

Sometimes damping-off diseases may not wait to attack until the seedlings are up. They hit the seeds as they are sprouting, and kill them before they get through the medium with which they were covered. Newly sprouted seedlings, especially those grown from unusually small seeds, are the most susceptible, but disease may occasionally also attack the husky seedlings of squash, watermelon, and even good-sized transplants, if conditions are right.

Pythium, Rhizoctonia, Fusarium, and Phytophthora are the most common damping-off diseases, but a grower may encounter several other root pathogens as well. Ordinarily a home grower doesn't need to be concerned about the proper name for the disease or diseases he discovers, because the treatment for all is similar. Commercial growers usually treat disease by blowing various gases and smokes through their greenhouses at night when no one is present. It is safer and more practical for small-scale gardeners to rely on sanitation and fungicides, and to be careful not to buy infected plants from other growers. Gift plants, we've found, present the greatest dilemma. To be on the safe side, we isolate them for a week or so and check them carefully before letting them in with the rest of our collection.

Even with the best of care, disease can strike your seedlings, and you should always be on guard. Early attention is extremely important. Your first inkling of an attack may come one day when you are proudly admiring your beautiful green seedlings, and suddenly notice that a few are beginning to wilt and fall over. The next inspection shows more casualties. Sometimes every plant in a large flat will collapse within twenty-four hours, though it usually happens over a longer period.

Fortunately, good fungicides are available that are excellent for helping to control disease. Captan is one of the most common. Subdue, Dexon, and Ben-Late (benomyl) are also in common use and readily available. The directions on the package should always be followed carefully.

Apply any fungicide with a fine spray, just after you have finished watering. If you have only a small number of seedlings, they can be easily sprayed with a mister like those on bottles of window cleaners. Beware of using insecticides and fungicides on food plants such as herbs and lettuce that might be broken off and nibbled by an unsuspecting visitor. The toxicity of most fungicides is considered to be low, but none of us feels comfortable enough about their potential danger to eat them. If you must spray, post warnings on your plants, and wait the time recommended on the package before eating.

A fungicide should always be considered a supplement to good growing practices, and never as a cover-up for neglecting any of them. Prevention of disease is always better for the plants than trying to find a solution after it has occurred.

To help prevent disease:

(1) Use a sterile soil mix, and place it in sterilized flats. (New flats, or old ones that have been washed in a Clorox solution are suitable.) Be careful, as you work, not to contaminate the media or seeds with infected tools or hands.

(2) Plant fresh seeds that will sprout rapidly and grow fast. Healthy plants resist disease better than weak, spindly ones.

(3) Don't overcrowd seedlings. Provide for plenty of air circulation around the plants.

(4) Water extremely carefully. Avoid cold water, dirty water, and over-watering.

(5) Provide the proper amount of heat and light. Fertilize carefully to avoid burning the tender roots.

TRANSPLANTING SEEDLINGS

Various soil mixtures may be used for transplants, and each grower has a favorite. I prefer a mixture of one-third loamy soil, one-third sand, and one-third of either shredded peat moss, homemade compost, or composted bark or leaves (all measurements by volume). To each wheelbarrow load (approximately four cubic feet, or a bit over three bushels), I add a quart of dried manure, one cup of bonemeal, and, for those plants that like a sweet soil, a cup of lime. Sometimes it is difficult to obtain dried manure, so I use a slow-acting fertilizer, such as Mag-Amp, instead.

Large growers use power mixers, cement mixers, or similar equipment for mixing their media. Small mixers are available for home growers from nursery supply houses and farm stores, or you may prefer to mix your potting soil by hand with a shovel or spading fork. Whatever the method, it is important to meld all the ingredients together thoroughly. If the mixing is done haphazardly, a small pot or flat may easily get too much or too little of some vital ingredient.

Mix in Fall

It is best to mix the soil in the fall and store it over the winter in a potting shed or greenhouse. It can be stored in bins under the benches, or in large plastic garbage cans. Whatever the weather, the soil is ready to use in early spring. By mixing it early and overwintering it, you give it a chance to "cure" and the ingredients become well blended. If you use Mag-Amp, or a similar slow-release chemical fertilizer, however, don't add it to the soil until the day you are ready to transplant. The nutrients begin to release as soon as they come in contact with moisture, so if the available fertilizer is added to the soil far in advance of transplanting, it could accumulate and be too strong for small plants, or if it is added too far ahead, it might all leach away. To save time and trouble, you may want to buy a

When transplanting, take up a bit of soil with each small plant.

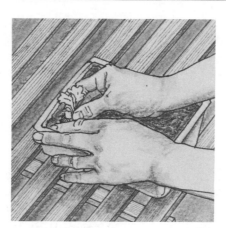

Set transplant slightly deeper than it was growing in flat.

Press the soil firmly around the roots to eliminate air pockets.

prepared soil mix if it is available at an acceptable price.

We use brown flats made of wood pulp for the transplants. A four by six-inch flat holds a half-dozen plants and one six by eight-inch holds a dozen. You may prefer peat pots of various sizes, divided containers, styrofoam cups, or any of a host of other products available. (See "Containers", page 8.)

True Leaves

The first leaves produced by your plants are the seed leaves. Transplant the seedlings as soon as they have grown their second set of leaves, the true leaves. If you do it sooner, there probably won't be enough root system to keep the plant alive, but if you do it later, the roots often will have become entangled and will be broken in the transplanting.

If you have guessed wrong about your planting date and the seedlings seem to be growing much too fast and are becoming leggy, you can check their growth somewhat by withholding fertilizer, water, or heat. Although it is possible to do this safely to some degree, it is risky business for the plants, and not recommended as a normal practice. It is better to transplant them and pinch back the tops later if they become too lank before you can safely put them outdoors.

If you are transplanting in a greenhouse, do it in the evening or on a cloudy day, so the plants aren't as likely to wilt. The bench of a potting shed is an ideal place to do your transplanting, if you have one.

Water the seedlings a few hours before you transplant them. Being full of water, they will stand the shock of transplanting better. Also, when the planting medium is wet, it will cling to the roots, and there will be less danger of root breakage.

I like to use a small plastic knife for the transplanting, but a little wooden pot label works well, too. Dig and separate the seedlings with care, trying to get a block of soil with the roots. The tiny root fibers that are so important to each plant's survival and good growth are very fragile. Dig only a few seedlings at a time, so they won't dry out. Transplant only the strong, healthy, perfect plants, and discard any weak, spindly ones.

Use Warm Soil

Be sure that the soil in your pots and flats is slightly warm, and moisten it with warm water. If you were to transplant young, tender seedlings into cold soil, and water them with cold water, you would ensure their failure.

You may prefer to use a dibble instead of the knife to make the holes in the soil. Gently insert the seedling in the hole in such a way that all its roots will be buried. Set it slightly deeper than it was in the seed flat, making sure it is evenly spaced from its companions, and standing up straight. Press the soil firmly around the roots, leaving no air pockets. Be sure that the soil is level over the surface of the flat or pot, so no indentations are left which might distribute the waterings unevenly.

Water Lightly

After the plants are all settled in their new homes, water the flats lightly with slightly warm water to which half the recommended amount (on the package) of liquid fertilizer has been added. If the greenhouse is hot and sunny, place them under a bench, or cover them with a single layer of newspaper to shade them until they recover from the shock of being transplanted.

Once every two weeks give the new plants liquid fertilizer, following the directions on the package. Remove any sick-looking plants, as soon as you notice them, and use a fungicide if there is any sign of damping off. Continue to check for insects, and spray if necessary.

Commercial growers often use a wetting agent on their flats to help the soil absorb the water better, and prevent it from hardening. Small growers usually prefer to cultivate their flats occasionally by dragging the pointed end of the dibble stick between the plants to break up the soil crust. If you cultivate, be careful not to disturb the roots.

We transplant many seedlings into pots rather than flats, because potted plants can be set directly into the ground without disturbing their roots, and consequently make much faster growth in the garden. We pot many vegetables, flowers, and most herbs in three-, four-, or five-inch pots, depending on how large we want them

SUGGESTED PLANTING SEQUENCE OF BEDDING FLATS FOR PLANTING OUTSIDE IN MAY

	Flowers	Vegetables
January	Begonias, Gloxinias Double Petunias	
February	Giant Dahlias Impatiens Pansies Seed Geraniums Petunias Tall Marigolds	Celery Onions
March	Asters Bachelor's Buttons Coleus Portulaca Snapdragons Zinnias	Eggplant Herbs Peppers Tomatoes
April	Alyssum Dwarf Marigolds Other annuals	Broccoli Cabbage Cauliflower Lettuce Melons Pumpkins Squash

to be when we set them out. Our hybrid tomato and pepper plants are transplanted, one to a six-inch or eight-inch round pot. When we set them in the garden in June, they are not only in bloom, but often are covered with small fruits.

PERENNIAL FLOWERS, HERBS & FOOD PLANTS

Landscaping styles come and go, but the perennial flower garden is always in vogue. A perennial bed enhances a country cottage or a fine estate and provides an everchanging mass of colors and shapes all season, year after year. Some perennials, such as daylilies, iris, and peonies, don't come true from seed, so they are propagated by division.

Among those most easily started from seed are Canterbury bells, columbine, English daisy, delphinium, forget-me-not, foxglove, hollyhock, iberis (candytuft), lobelia, lupine, lythrum, pansy, pinks, pyrethrum, shasta daisy, snow-in-summer, sweet william, and viola.

Perennial herb and food plants that can be started from seed include asparagus, catnip, peppermint, rosemary (tender perennial), rhubarb (common kinds), spearmint, sage, thyme, wormwood, and many others.

Buying and Collecting Perennial Seeds

The perennial sections of seed catalogs are tempting and offer many wonderful varieties. They often can be confusing, however, if you are unfamiliar with plant propagation, because some list the seeds of many varieties that are best started in other ways. Bleeding heart, for instance, although it can be started from seeds, will not grow easily, and it is much better to propagate it by division, or cuttings. Oriental poppies and perennial phlox grow easily from seed, but are likely to produce mostly plants that don't have very exciting blooms. Refer to the list of herbaceous plants in the back of this book which states the best way to start each perennial.

Gathering Seeds

You may want to gather part of your own seed. When I first started in the nursery business, I collected many of the seeds I used. I didn't save any from annual flowers, because far better ones could be easily bought, but I successfully started those collected from delphinium, English daisies, forget-me-nots, hollyhocks, lupines, pansies, sweet william, violas, and many others.

I had mixed feelings about collecting them, because all the garden books said that perennial flowers should be picked and should never be allowed to go to seed. Producing seed does weaken a plant, makes it bloom for a shorter period, and often shortens its life as well, but since one plant produced all the seed we usually needed, we willingly made that sacrifice.

Seed may be gathered without guilt from the biennials. They are likely to bloom only once and die shortly afterwards. If unpicked, they quite often spread their seed around the garden and reproduce themselves.

Some of the named varieties of hybrid perennials don't produce 100 percent like the parents, naturally. Seeds of chrysanthemum, for example, may produce plants showing a wide variation.

Let Them Dry

If you collect perennial or biennial seeds, leave them on the plant until they are thoroughly ripe and dried almost to the point at which the seeds will scatter. Remove them and let them dry for a few days longer in a warm place like a greenhouse. Then shake them from the pod, clean out the impurities, and either plant them or store them for later planting.

Some perennial seeds may be planted directly in the ground where you want them to grow, but most should be started in prepared outdoor seedbeds, or sown in flats in the house or greenhouse.

Perennial and biennial seed can be planted outdoors almost any time the ground can be worked easily. Early spring, mid-summer, or late fall are the best times. Seeds planted late in the summer may grow well, but not have time to develop enough roots to stand the winter without protection. If you must plant seeds then, give the tender seedlings protection for the winter. Plant them in a cold frame, if possible, and cover them with evergreen boughs. Seeds planted just before winter will often lie dormant and sprout the following spring.

Preparing the Soil

Outdoor seedbeds should be of light, loamy soil and well tilled to a depth of at least ten inches, with all weed and grass roots eliminated. The soil should be supplied with humus and nutrients, just as if you were planting a vegetable garden. Though it is not absolutely necessary, before planting it is beneficial if the bed is first sterilized with Vapam or a similar chemical and then thoroughly aerated so none of the fumes remains. This treatment not only gets rid of the pathogens in the soil, but also eliminates weed seeds.

Plant the seeds at a depth of three or four times their diameter. If planted in summer, most will germinate better if they are started under a light shade. A shaded cold frame is an ideal spot. It is also helpful if the beds can be screened, to

---

HARDENING OFF PLANTS

When seedlings have been transplanted into flats or other containers, they should be prepared for life in the outdoors.

This is called hardening off.

It means preparing plants to endure and grow despite changes in temperatures, variations in the amount of moisture, and winds.

Many gardeners do this in a cold frame, keeping it closed at first, then gradually exposing the plants to outside conditions for a greater number of hours each day.

Another method is to move the plants daily from the house to a shaded, protected area. They gradually can be exposed to sunlight and winds, and finally left out overnight.

Careful hardening off means plants are prepared for the more severe conditions in the garden, and thus will grow there without setbacks.

protect them from the mice and birds who like to eat the seeds. It is easy to start small amounts of seeds outdoors in the summer by planting them in flats, using an artificial soil mix, just as you would start annuals. The flats can either be placed on the north side of a building, or shade can be provided in some other way.

Perennials may be started inside in flats in late winter or spring, just as annuals are, and unless the sun is very bright, no shade is necessary then. The temperature required by perennials for germination is not as high as that needed by annuals; 60 to 65 degrees F. is about right for most of them.

Many perennials will bloom the same year if planted inside early. Except for the difference in temperature needed for germination, the culture is much the same as for annuals. Although most perennials are not as subject to damping off as are annuals, the same sanitation and watering practices should be followed.

Self-Seeding Perennials

Some perennials do an excellent job of planting their own seeds. The more common kinds of lupines, foxgloves, English daisies, violas, and others self-sow so successfully that the plants often escape the garden and become nearly wild. The best varieties of perennials need more care, though, both in propagation and culture.

DECIDUOUS TREES & SHRUBS

Most gardeners never consider starting trees and shrubs from seed. These plants seem to be in a different class from asters and swiss chard, and perhaps people conclude that, because of their large size, it would take too many years to grow them to maturity. Growing your own landscape plants is an interesting hobby, however, and it may be the only way to get certain unusual varieties. It is also practical if you need a lot of trees for some reason, such as a long hedge, windbreak, or woodlot.

You may want to plant seeds for rootstocks, too. Most fruit trees, tea roses, many nut trees, and some named varieties of shade and flowering trees are grown by grafting or budding the desired variety upon a seedling rootstock. We grow lots of apple seedlings from the seeds left in the pulp after our fall cider making. These seeds are planted, pulp and all, in beds, and they sprout the following spring. A year later we transplant them into pots or rows, to be bud-grafted the same summer.

You may find you can gather most of the seeds you want from the trees and shrubs growing in your own area. No one is likely to care if you collect seeds from desirable trees growing along streets or in parks or cemeteries, or from your neighbors, whereas someone might object heartily if you were to dig shoots or take cuttings. In most areas you will find many native trees and shrubs that are worth propagating for their beauty, their ability to attract birds, their shade, or other reasons.

Not every seed you find will produce a tree or shrub exactly like the one from which you gathered it. The seeds of old-fashioned lilacs, wild roses, and most native trees and shrubs have standardized themselves and will produce plants much like their parents, but a seedling grown from a named variety, such as a Ludwig Spaeth lilac, is not likely to do this. Still, you may enjoy experimenting with French lilac seeds, as well as the seeds of named varieties of azaleas, rhododendron, holly, dogwood, flowering crabs, spireas, roses, fruit trees, and similar plants. Some of the results may be quite good, and you may find yourself growing a superior new variety.

It may be tricky to find out when seeds are ripe and to pick them at the right time. Birds and animals know this instinctively, but you may have to observe carefully to get them just as they are ready to fall. Seeds may come in pods (catalpa), nuts (oak), flowering raspberries, fruits (hawthorn), and other forms.

If you plan to propagate large numbers of trees or shrubs, you will probably need to order some tree seed by mail. Experimenting is always fun, but check with a tree and shrub encyclopedia to see which varieties do well in your area. The appendix lists the names of companies that sell seeds of woody plants.

PLANTING THE SEEDS

Obviously Nature has great success planting trees and woody shrubs, so we simply imitate her and make a few improvements here and there. Improvements are necessary because Nature is far more patient than we are, and we also want a large percentage of the seeds we plant to grow.

In nature, seeds are planted as soon as they are ripe, and we should do the same whenever possible. In fact, a few highly perishable seeds, such as those of certain maples, elms, birches, and willows, should be planted as soon as they fall from the trees. They need no drying or curing. In sugar maple and similar seeds, the embryo forms early and becomes well developed inside the seed before winter. It starts growing immediately as soon as the weather warms the following spring.

Sometimes it is not possible to plant seeds as soon as they are collected, so read the directions for each variety in the back section of the book to find out about any special seed treatment, such as stratification, that may be necessary or beneficial before planting. Certain plants are quite demanding, so you may want to leave the growing of these to the experts.

The propagation procedure for tree and shrub seeds is much like that of starting perennials. The same directions for planting apply, as do the cautions about watering, sanitation, and other care. Tree and shrub seeds are not usually planted indoors. They may be started in flats in a cool greenhouse, however. Seed may be planted outdoors in flats, too, or in open beds or cold frames. Outdoor seed beds can be of any size, but if you plan to start a large number of plants, make the bed about five feet wide, and as long as you want it. As with perennials, the bed should consist of well-tilled, rich, light, loamy soil, and if the bed is slightly raised, it will give the new plants more root area for growing. Most deciduous seedlings don't need shade in order to sprout, but a light shade or thin mulch of vermiculite or sifted compost is useful to help keep the soil moist.

How to Plant

The seeds can either be planted in rows or broadcast around the bed. Don't plant them too thick since they won't grow as well, and you will have to thin them later. Cover them with a thin layer of soil, sand, or perlite, and water them thoroughly. Finally, scatter in a few moth balls to discourage mice and squirrels.

Never let the upper layer of soil dry out while the seeds are sprouting and growing. It may be necessary to water the young plants every day during dry periods until their roots have grown deep enough to reach moisture. Check the seedlings frequently for disease and insect damage, and give them a feeding or two of a liquid fertilizer, but stop all feeding by mid-July. Later than that will encourage too much late-season growth.

The seeds of different varieties of woody plants germinate at different speeds, so be patient. Most take far longer than do perennials. Some, such as those of lilacs, may take a year to germinate if you plant them in the spring. Others may be up in a few weeks.

In this seedbed, plant seeds thinly so you won't have to thin later.

Use a thin layer of sand, soil, or perlite, then water thoroughly.

Finally, scatter a few mothballs. They will discourage small animals.

Transplanting Seedlings

Deciduous seedlings grow throughout most of the summer, so if all goes well, the plants in your seed bed should be fairly good-sized by the end of the first growing season, and many will be ready to transplant the following spring. If they then have a heavy, well-formed root system, the plants can be set out where you want them to grow permanently, or they may be planted in nursery rows to grow larger, or transplanted into large pots.

If the root system is weak, the seedlings should either be kept in the seed bed a second year, or transplanted into another bed, spaced six to eight inches apart, and allowed to grow an additional year. Prepare the transplant bed as thoroughly as the seed bed, continue a good program of weed control and fertilization, and never let the plants dry out. The secret of many a gardener's success is to give the plants water when they need it.

Certain plants, including most trees that produce nuts, such as beech, butternuts, oaks, and walnuts, should be transplanted before they develop their characteristically long taproot. Cutting off or bending the taproot weakens the tree, and may even kill it. Digging and moving this type of tree is far easier and more successful if done before the tree gets more than four or five feet tall.

STRATIFICATION OF SEEDS

Certain seeds need a moist, chilling period in a dark place before they will sprout. If you were to gather such a seed when it appeared to be ripe, and plant it immediately in a warm greenhouse, it would probably not grow; nor would it germinate readily if it were stored over the winter in a cool, dry closet. These seeds must be placed between layers of moist soil or other material and kept cool in a process called stratification. This stratification softens the seed coat so water can pass through it, permitting the seed to develop into a tiny embryo. As soon as the soil is warmed in the spring, the embryo breaks through the seed coat and forms roots and leaves.

Nature takes care of the process very nicely. Seeds fall from the trees to the ground and are either covered by leaves or planted by squirrels where they remain protected by the cool, wet soil until spring. Whenever we can, we should imitate Nature by planting the seeds outdoors as soon as they are ready.

Sometimes immediate planting is impossible, so we must stratify the seeds in an artificial manner. An easy way to do it is to place layers of seed between layers of moist but not wet sand, or a mix of sand and sphagnum or peat moss, in plastic cans or wooden boxes. If the seeds are small, spread them between two sheets of cheesecloth or fine plastic screening, to make it easier to separate them from the sand mixture later. Bury

To stratify seeds, place layers of them between layers of sand in a plastic can, then bury can under leaves.

the container in layers of leaves in a hole in the ground outdoors, or keep it in a cold rootcellar for the winter. You can stratify small amounts of seeds in glass jars or plastic bags in a refrigerator (not in the freezer). If you store them outdoors, include a few mothballs with the seeds if there is danger of mice, rat, or squirrel infestation.

Seeds should be stratified only during the winter previous to spring planting, and those that are to be held in storage for a year or more should not be stratified before storing.

SEEDS THAT BENEFIT FROM STRATIFICATION

Apple (Malus)
Apricot (Prunus)
Ash (Fraxinus)
Barberry (Berberis)
Beech (Fagus)
Bittersweet (Celastrus)
Boxwood (Buxus)
Cherry (Prunus)
Corktree (Phellodendron)
Cotoneaster (Cotoneaster)
Cranberry Bush (Viburnum)
Dogwood (Cornus)
Firethorn (Pyracantha)
Flowering Quince (Chaenomeles)

Fringetree (Chionatnthus)
Ginkgo (Ginko)
Hackberry (Celtis)
Holly (Ilex)
Honey locust (Gleditsia)
Honeysuckle (Lonicera)
Hornbeam (Carpinus)
Horse chestnut (Aesculus)
Juniper (Juniperus)
Linden (Tilia)
Locust (Robinia)
Magnolia (Magnolia)
Maple (Acer) Varieties that ripen
        in the fall only.

Nandina (Nandina)
Pawpaw (Asimina)
Peach (Prunus)
Pear (Pyrus)
Persimmon (Diospyros)
Plum (Prunus)
Privet (Ligustrum)
Russian Olive (Elaeagnus)
Shadbush (Amelanchier)
Sweet gum (Liquidamber)
Tulip tree (Liriodendron)
Yellowwood (Cladrastis)
Yew (Taxus)

DORMANCY

Different seeds vary greatly in the time it takes for them to sprout after planting. It takes patience to wait for morning glory, sweet peas, celery, and asparagus seeds to germinate, while radishes and marigolds come up in two days if conditions are right.

Survival Mechanisms

There are many reasons for seed dormancy. To assure their continued existence, some seeds apparently have developed methods of surviving for long periods under adverse circumstances, and they germinate only when conditions are right for growth. The germination appears to be regulated by various mechanisms including a hard seed covering that won't either let moisture in or allow the embryo to expand, a chemical growth inhibitor, and others.

Although it often seems as if our vegetable and flower seeds take a long time to germinate, certain tree and shrub seeds are much more deliberate. Many have hard coats that take years to soften naturally. The Brazil nut reputedly requires six years before the shell is softened, and two additional years to sprout. Locust, pea shrub, Kentucky coffee tree, acacia, and other seeds also have extremely hard coats.

Weaken the Seed Coat

Commercial growers sometimes use machines to crack hard seed coatings. They also treat seeds with concentrated sulfuric acid to soften the coating and encourage faster germination, a process called acid scarification.

Rather than experiment with a potentially dangerous acid treatment, home propagators sometimes make a nick barely through the coat of hard-to-crack seeds with a small file or bit of sandpaper. Large seeds can be touched gently to an emery wheel to open up their hard shell slightly to let moisture come through.

The Easiest Method

Soaking seeds in slightly warm water before planting is the easiest and most frequently used method for softening coatings, and this works well with a great many. In addition to hard-shelled seeds like morning glories, such ordinary ones as peas, corn, and beans are often soaked in water that is barely warm for a few hours before planting, to speed up germination by a few days. The second section of this book lists treatments that some seeds may need prior to planting. Check the varieties you plan to grow.

EVERGREENS FROM SEED

The term "evergreen" is confusing because it is misleading. Sometimes the word conifer (cone-bearing) is used instead of evergreen, but that doesn't quite fit either. Larches are conifers, but they are deciduous, not evergreen. Yews and junipers are evergreen, but their seeds are contained in berries, not cones. The expression "plants with needles" doesn't help either, since arborvitae and cypress don't have needles, but they are evergreen conifers.

Also, there is an entire class of evergreens, such as holly, azaleas, and laurel, that are neither conifers nor needle bearers. They are the broadleafs, some of which are "evergreen" everywhere, and others that are deciduous in the North and evergreen in the South.

The method of growing broadleaf seedlings, junipers, and yews is the same as that for deciduous trees and shrubs. The term evergreen, as used here, refers only to the conifers.

Growing evergreens from seed can be a very worthwhile project. You can produce your own landscape shrubbery at very little cost or grow large numbers of trees for woodlots, Christmas trees, and reforestation.

Many people grow a variety of evergreens in the same bed to meet several needs.

Buying Seed

Several companies listed in the appendix offer a large assortment of evergreen seeds. If you want to know the seed source, the company will usually supply that information, which is useful if you live in northern Maine and don't want to plant seed that was gathered in Arizona.

It is tricky to gauge exactly the amount of seed to buy. Although there are a lot of seeds in one ounce, not all of them are viable. Usually it is best to start with a small packet of seed the first time, and expand as you become more familiar with the process.

I like to order our tree seed in late summer because some varieties are always in short supply, and by ordering early, I am more assured of getting the best choice. Most seeds are not gathered until fall, so they are not likely to arrive until late fall, and sometimes come in different shipments. Occasionally some arrive so late we must store them for spring planting.

Collecting

I collect as many seeds as possible. Seeds gathered from native trees are fresh, and often germinate better than those from other areas. The plants grown from them will be acclimated to the region, too. Evergreens usually produce seeds only every other year or even less often, so we must be prepared to collect them those years when there is a good crop.

You don't have to live on the edge of the North Woods to gather seeds. You can probably collect small amounts from ornamental evergreens in your home landscape and public parks. Mugho pines, blue spruce, arborvitae, yews, and many others produce seeds that are well worth planting. Named varieties of evergreens, such as pyramid arborvitae, which don't come true from seeds, often produce them, and although the seedlings of such plants may not grow into specimens that resemble the parent tree, they may develop into something interesting.

It is tricky to know when the seeds are ripe and ready to gather. I have found that watching the squirrels is the most reliable way, since they always seem to know exactly. Collecting seeds from tall trees is more difficult for people than squirrels but if you are interested in growing forest trees, sometimes you can find a pulp or lumber-cutting operation going on in late summer or early fall. It is easy to pick the sticky seeds from the tops of fallen trees, and you can often gather all you want in a few minutes.

Conifer seeds are enclosed in cones that vary in size from the tiny half-inch eastern hemlock cones to those of the sugar pine, a western tree which often has cones of twenty inches or more in length. Be sure to wear a hard hat when you are collecting these.

Most cones hang downward from the tree, and as they dry, they open and release their seeds. The empty cones later fall to the ground. Fir cones, however, stand upright like little candles at the tops of the trees, and disintegrate there, and the seeds and bits of cones flutter down together.

If you want seeds, gather the cones while they are fat and ripe but still tight, slightly green, and covered with pitch. Gloves are convenient, but I

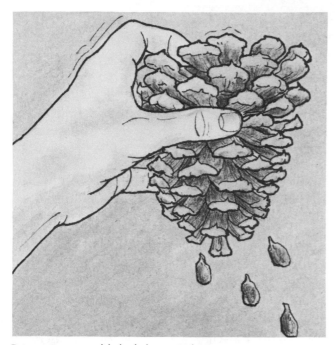

Dry cones—and label them. When cones pop open, it is easy to shake out and collect their many seeds.

prefer to work with my bare hands even though it takes lots of kerosene to clean them later on. Put the cones in bags, and be sure to label them if you gather more than one variety at a time. I think I can identify a red spruce cone and distinguish it from that of a white spruce, but when I get home, I'm not so positive.

Dry Promptly

Prompt drying is important, or the cones will rot. A sunny greenhouse, an airy, warm shed, or outdoors in the open sun are good places to dry cones. Spread them out one layer deep, and place them so they will not crowd each other, on a wire screen that is elevated a few inches to allow the air to circulate beneath them. Don't let rain fall on them or allow heavy dews to dampen them because any excess moisture will cause them to rot.

As the cones dry they pop open, and you can shake out the seeds. Each seed has a wing on it which enables it to float for a long distance when it is released from a tall tree. If you buy seed you will find that the wings have been removed, making the seed much neater and cleaner. It is not necessary to remove them, however, since the seeds grow just as well if the wings are left on.

Evergreen seeds sprout best if planted in the fall, as soon as the seeds are separated from the cones. They will germinate as soon as the soil warms in the spring. If you buy seeds, however, it may be necessary to store them until spring; and if you collect them, you may want to save seeds during those years the trees produce so you will have some to plant in the years they have none. Keep the seeds in a plastic bag in a cool, dry place. Although the storage life of different varieties varies, most will remain viable for two to four years under good storage conditions. Firs are an exception and do not germinate well after the seeds are a year or two old.

If you are growing only a few seeds, you may plant them in pots or flats indoors in the house or greenhouse, or in a shaded cold frame outdoors. Follow the directions for planting perennials, but keep them out of direct sun. Evergreens, like perennials, will germinate at a lower temperature than annuals, but their moisture and other care will be about the same.

The Seed Bed

Outdoor beds are more practical for planting large amounts of seeds. The procedure is similar to that of starting deciduous and shrub seeds outdoors, except that evergreens need the kind of shade they would get if they were sprouting in a forest. They, too, thrive in a sandy loam, so if your soil is heavy, mix sand with it before planting. Till the soil thoroughly, add well-rotted manure or compost, and then till it again to mix all the ingredients thoroughly.

Because evergreen seedlings are very susceptible to damping off, we sterilize the bed with Vapam before planting the seeds. The chemical not only gets rid of harmful soil organisms, but also kills weed seeds and roots. We follow the directions on the can, let it set the recommended time, and then till the bed thoroughly once more to release all the gas. As the final step of preparation, we shape the soil into raised beds that are approximately five feet wide, eight to ten inches tall, and the necessary length.

To plant the seeds, scatter them across the top of the beds, keeping in mind that, although not every seed will grow, those that do germinate should have room to grow well.

Next, cover the bed with about a quarter-inch of sand, and spread moth balls generously over the top to confuse the mice and squirrels. I pile several thicknesses of evergreen boughs on top of the sand, so that winter's freezing and thawing won't heave the seeds out of the ground. If evergreen boughs are hard to come by, leaves or clean straw are good substitutes.

When the temperature begins to rise in the spring, remove the mulch and build a shade frame over the bed. My system is to drive posts (about three feet high) into the ground every six feet, along each side the bed. I nail a narrow board lengthwise along them about eighteen inches above the ground, and lay short boards about two feet apart across the width of the bed over the two nailed on boards. Then I pile on the boards all the brush that had mulched the bed during the winter, to make a heavy shade. Other neater appearing materials, such as plastic shade cloth, snow fence, or wood lath, can be used if you like.

The germination time varies, but most seeds sprout within a few weeks, covering the bed with a mass of light green. The time after they have just sprouted until they are an inch and a half tall is the most critical period for the seedlings because they are most susceptible to disease then. Because the shade blocks out the sunlight and breezes that usually control damping off, the virus, once started, can multiply rapidly among the cool moist seedlings and quickly kill them. To help prevent disease, as soon as the seedlings are up, spray the bed every week with Captan or another good fungicide. Thin out the seedlings so they don't touch each other. They need room to grow, and they resist disease better if they are not too crowded. Remove any weeds, also. Birds may present a problem, too, and some screening on the sides of the bed may be necessary to keep them from pulling the young seedlings.

Water the bed if the weather is dry, and give the seedlings two or three doses of liquid fertilizer, such as Rapid-Gro, a week apart. Stop all feeding before mid-summer. You will find none of these chores is easy when you have to work under a heavy shade cover, but your care will reward you with healthy plants.

The evergreen boughs begin to dry and drop their needles by mid-summer, letting additional light fall on the bed, which is beneficial to the trees at this stage of their growth. The fallen needles also make a nice mulch between the seedlings, creating an environment similar to what they would have if they were growing in the woods. If you have used another kind of shade, remove part of it in mid-summer, or replace it with another that lets in a bit more light.

Depending on their variety and growing conditions, the seedlings will grow from one to eight inches by late summer. By then they will have stopped growing, so you should remove all the shade.

The following spring, put a light shade over hemlock seedlings and any fir seedlings that are less than three inches tall. Taller firs, as well as spruces, pines, and arborvitae (native white cedar), need no shade the second spring. Sprinkle dry manure or 5-10-10 between the seedlings one day in early spring when they are dry, so the fertilizer won't stick to the trees.

Scatter seeds on raised bed in fall, cover with sand, drop on moth balls, and cover with boughs.

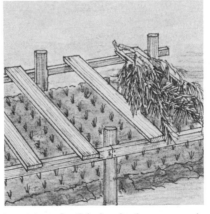

In spring, build shade frame, and pile on the same boughs, or use any other shading materials.

The third spring, move all trees, small and large, to a transplant bed for two more years of growth.

Transplanting

The third spring, the young trees should be moved to a transplant bed, even if some are still small. If you were to leave them in the seed bed for another year, they would be overcrowded, grow tall and straggly, and develop very few roots. Evergreens may be moved in very early fall, their second year in the seedling bed, if they are kept watered and heavily shaded. If planted at this time, they often need only one year in the transplant bed.

A transplant stick is very handy for setting out large numbers of seedlings. (See box.) Whether you are transplanting your own seedlings or some you have bought from a commercial nursery, it is a big timesaver and helps you space the trees evenly so each has enough room. Soil in the transplant bed should be as well prepared as that in the seed bed. It should be from four to six feet wide, either at ground level or formed into a raised bed.

Prefer Sunlight

Most evergreens prefer full sun, so usually no shade is needed for evergreen transplant beds. Hemlocks are an exception, however. If they are

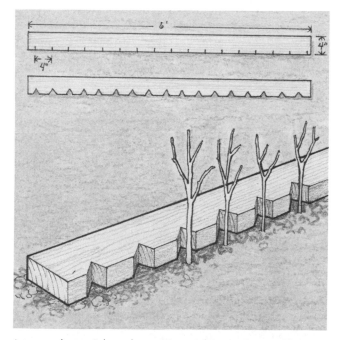

A transplant stick such as this can be made quickly and easily (see next page), and guarantees that you will space trees correctly, and that rows will be straight.

less than six inches tall, they are too frail to endure bright sunlight all day, and should be shaded for one more year. Even as they get older, hemlocks can stand shade well, and thrive under conditions too shady for most other trees.

Unless growth is exceptionally good, allow the trees to grow in the transplant bed for two years. The first year the trees usually grow very little above the ground, but their root systems develop a great deal. Liquid feeding with Rapid-Gro or a similar chemical or organic fertilizer, at weekly intervals several times in early summer, gives them an additional boost. The second year the tops should grow rapidly and often double in size.

The third spring they should be planted out in the field where they are to grow permanently or transplanted into nursery rows or pots to grow larger and be sold later as small trees. If they are left more than two years in the transplant bed, they will become stunted.

Buying Seedlings

If you wish to avoid the work and time required to start your own evergreen and deciduous seedlings, you can buy them from mail order nurseries. Some state forest services operate nurseries, too, and occasionally sell seedlings to individuals. Consult with your area extension office to see if such a service is available in your state. Seedlings bought through such services in large numbers are surprisingly inexpensive.

Nursery catalogs describe their trees by variety, the size (usually in inches), and a set of figures such as (2-0) or (2-2). The first number inside the set of parentheses refers to the number of years the seedling has spent in the seed bed, and the second, to the number of years in the transplant bed. (2-2) would mean that the tree was four years old, and had spent two years in a seed bed and two in a transplant bed. (2-2-2) would indicate that a large tree had been transplanted twice and spent two years in each bed.

If you want seedlings for transplanting, you would order the (2-0) size. For field planting, however, you would probably want the larger transplant (2-2).

If you buy plants, put them in a cool shed or basement as soon as they arrive. Keep them out of the sun and wind, and soak their burlap coverings with water.

THE TRANSPLANT STICK

A transplant stick makes it possible for one person to transplant three or four thousand seedlings in one day, and be certain they are all the same distance apart. I make our beds six feet wide, to get the most seedlings in a small area, but some like them narrower.

A board about an inch thick and four inches wide is a convenient size. Cut it two inches longer than the width of your beds, and beginning one inch from an end, mark it off in four-inch spaces. Then cut a V-shaped notch into the board at every mark.

Beginning at one end, place it across the transplant bed, the straight edge of the board toward the area to be planted. With a square-bladed shovel cut a trench in the soil, close to the board's straight edge and just wide and deep enough to hold the roots of the seedlings. if the soil is unusually dry, fill the trench with water.

Next, turn the board around so the notched side is in front, and place a seedling in each notch with its roots in the trench. When the first row is complete, with your hand or the shovel, pack the loose soil firmly around the roots of the trees, making sure that the seedlings are standing straight, and no air pockets are left about their roots. Then remove the board, being careful not to pull out the newly set trees, and move it ahead about six inches to make a new row. If you have a six-foot bed, you can plant twenty-five trees in only a few minutes.

Choose a cloudy day with no wind for the transplanting, if possible, and carry the trees to the planting site with their roots in a pail of water. Think of them as trout, and never let them dry out. When transplanting, stop every few rows, and water the trees with a hose or sprinkling can.

Place transplant stick across bed, to mark site of first row of seedlings.

Dig a trench wide and deep enough to contain the roots of the seedlings.

Flip over transplant stick, then position it on the edge of the trench.

Place one seedling in each notch, positioning the roots in the trench.

Pack loose soil around the roots; make certain there are no air pockets.

Move transplant stick six inches down bed and repeat the planting process.

GROWING YOUR OWN VEGETABLE & FLOWER SEED

During the years of the Great Depression we saved many of our own seeds, even though a packet cost only a nickel or dime. We dried them carefully, stored them in a cool, dark closet, and planted them in the spring. The home-grown seeds germinated faster than many of those we bought, and usually the vegetables matured a week or so earlier, which was a bonus in our short growing season.

We stopped growing our own seeds primarily because it limited our selection of vegetable varieties. When we saved seeds we couldn't raise hybrid corn or any other hybrid vegetable because the second generation of any hybrid is likely to be quite inferior to the parent. Furthermore, we could grow only one variety of each vegetable. If we grew more than one kind, the bees would cross-pollinate them, and the resulting seed produced a kind of corn, tomato, or whatever, that was quite different from what we had hoped for. Although the "squmpkins" that resulted when we planted the seeds that resulted from a pumpkin-squash summer romance were not too inferior in flavor, their appearance was weird. Not only that, our carrots were often pollinated by Queen Anne's lace or other wild carrots, resulting in ruined seed. Consequently, except for growing one variety of beans for drying, we now leave seed growing to the specialists.

If you are more adventuresome than we, and want to grow some of your own seed, make certain that you and all your neighbors within 600 feet grow the same variety of each vegetable or flower from which you plan to save seed, and of course, none of them should be hybrids.

Plants that mature their crops in a short season produce seed easily. Lettuce, radish, and spinach "go to seed" by mid-summer. Vegetables that produce a crop which is also its seed, such as peas, corn, and beans, are easy to grow, but getting the seed ripe enough to harvest and store takes a week or two longer than growing it to eat.

In warm climates, most vegetables bloom and produce seed the same year, but in areas with short growing seasons, some, such as carrots, beets, turnips, chard, parsnips, and members of the cabbage family, usually require two years. In the spring, some of these roots that have over-wintered outdoors under a mulch or inside in a root cellar, can be planted in the garden. They will bloom in mid-summer and bear seed in much the same manner as lettuce and radishes do the first year.

Harvesting

Sometimes it is hard to know for sure when a seed is ripe and ready for harvest. Pick off the kinds of seeds that hang on the plant, such as those of lettuce and parsnip, as they begin to fall off by themselves.

The "fruit" vegetables, including cucumbers, tomatoes, peppers, pumpkins, and melons, must be completely ripe but not rotten before picking. Separate the seeds from the pulp, and dry them carefully. One easy way to do this is to spread the seeds on paper towels, leaving as little pulp as possible. Keep them in a dry, warm area until they are completely dry. They will be stuck hard to the towel. Fold the paper carefully and store it in a large paper envelope in a cool, dry place until you are ready to plant the seeds the following spring. At that time, tear off little sections of the towel, plant two or three seeds in a pot of Pro-mix or other planting media, and pinch off all but one of them after they start to grow. Or, you may lay the entire towel, with the seeds still sticking to it, seed-side up, on a flat of Pro-Mix or other planting media, and cover it with a thin layer of perlite. The towel will deteriorate as the seeds grow.

Seeds that grow in pods, such as radishes, peas, and beans, should ripen enough to rattle around in the dry pod before you pick them. The plants can be pulled at that time, hung up to dry further, and the seeds shelled after they are thoroughly dry. Store them in a paper or cloth bag until spring planting time.

Corn should be ripened on the stalk a bit past the eating stage. Then the stalks should be cut to the ground and stood upright in shocks in the field. After a few weeks of "after ripening," the corn should be husked and dried on wire mesh or screens in a warm, well-ventilated garage, attic, or greenhouse until the kernels are hard enough so you can shell them easily. Corn, too, should be stored in paper or cloth bags.

If you dry seeds in flats or on sheets of plastic,

stir or turn them frequently so they won't rot before drying.

Storing Seeds

The best advice I can give you is to store seeds for no longer than one season. Always use fresh seeds so you will be certain they will be viable. Never order more than you intend to plant, and if you order too many, throw away what you don't use each year. Having said that, in honesty there are many occasions, whether we buy our seed or save it, that we want to store it for more than one season.

Although all seeds are perishable, they vary greatly in the length of their storage life. Most of them, even those with short shelf life, keep much better if they are sealed in jars or plastic bags and stored in a cool, dry place. Peas, corn, and beans, because of their possible moisture content, should be stored in cloth or paper bags to reduce the chance of rotting. Most flower and vegetable seeds keep for a year or so in a cool, dry place, even if they are stored in only their paper envelopes. Kitchens, hot attics, greenhouses, or damp basements are not good places to store seeds.

Always write the date on each seed packet. It is easy to forget whether seed is one year old or has been stored for five years. Although old seeds may be viable, they often germinate slowly, and grow at a slower rate than fresh ones, a characteristic not appreciated by those of us who live where growing seasons are short. Slow-growing seeds are also more likely than fast-growing ones to succumb to one of the damping-off diseases.

We store those seeds we keep in small amounts in a 5x8-inch metal file box. File divisions are labeled, so all the varieties of the same species are together.

Although the viability of most seeds lessens as time goes on, certain weed seeds, such as wild mustard, sometimes lie dormant for years, and sprout vigorously whenever the ground is cultivated. As a rule, farm seeds such as grasses and grains, stored in an ordinary shed, remain viable for one to five years. Most vegetable seeds can be kept for three to four years, but onions, sweet corn, and beets are reliably viable for only one or two years. Cantaloupe, watermelon, and cucumber seeds keep well for up to five years.

Most flower seeds, both annual and perennial, stay viable for several years if they are sealed in a plastic bag in a cool, dry place. Hybrid delphinium are an exception, and they are rather perishable. They should be planted within a year of harvest. We keep our delphinium seeds in the refrigerator, sealed in a plastic bag.

Testing Viability

Whenever you have flower and vegetable seeds that you feel are not fresh, test their viability before wasting your time and space in planting. Count out a certain number of seeds and plant them in a flat in late winter in a warm place. Then check, after a week or so, to see how many have sprouted.

The old, familiar "Rag Doll" method works well, too. Spread a number of seeds out on a moistened cloth towel or on several sheets of moist newspapers. Roll them up, tie them together with a string, and keep them slightly moist at about 70° F. Check in a week, and continue to check, to see what percent germinate and how long it takes.

Tree and shrub seeds vary even more than other seeds in their storage life. As I mentioned before, some of those that drop from the plant in early summer start to sprout almost immediately. These types, including the willows, elms, poplars, and most native maples, are difficult to hold in storage, even for a short time. Other varieties that should be planted as soon after they fall as possible include the alder, shadbush, false cypress, spice bush, liquidamber, magnolia, bayberry, Oregon grape, hop hornbeam, potentilla, sassafras, spirea, yew, and larch, as well as many nuts. In contrast, certain hard-shelled seeds, Russian olive and linden for example, may lie dormant for twenty years or more and still germinate, without any special storage conditions. In general, the seeds of most evergreens and deciduous plants, other than those with hard shells, have a viable life of from one to four years.

Although most tree and shrub seeds prefer a cool storage atmosphere, not all like it dry. Moist storage conditions are desirable for the seeds of maple, citrus trees, hop hornbeam, loquat, tupulo, and most of the nut trees, including oak, hickory, beech, walnut, chestnut, and filbert. Since it is difficult to provide ideal conditions for each kind of seed, plant as soon as the seed is ripe, whenever possible.

THE JOY OF POLLINATION

I was surprised, in my youth, to learn that nearly all plants have blossoms, even though the flowers of a pine tree or those on timothy hay don't look much like rose blooms. It was even more unbelievable to find that in order to produce seeds, each of the trillions of flowers over the countryside had to be pollinated in some way.

The flowerheads of plants have female organs (the pistil) and male organs (the stamens, which contain a yellow dust, pollen). Male and female organs may both exist in the same flower (bisexual), described as "perfect"; in separate flowers on the same plant (monoecious), described as "imperfect"; or on different plants (dioecious). Most fruits, and nearly all common flowers have perfect blooms.

Corn is a good example of an imperfect plant. It has male flowers (tassels) at the top, and the females (silks) further down. One grain of pollen from the tassel must hit each silk of the same or an adjoining plant and travel through it to the ovum, to produce each kernel of corn. When pollination is imperfect, your ear of corn at the dinner table will not be fully developed.

Holly, bittersweet, yews, bayberry, alpine cur-rants, and many others are dioecious. Date palms, too, are either male or female. Over 2,000 years ago Herodotus wrote of an annual ceremony of the Egyptians in which they waved the flower cluster of one date palm over the flowers of another. As a result, only the dates thus treated bore fruit. Apparently no one suspected at that time how it happened and the priests no doubt got credit for their miraculous powers. Even as late as the eighteenth century it was difficult for Europe and America to accept the fact that innocent little flowers were involved in anything as unpuritanical as sex.

If you buy dioecious plants such as holly or bittersweet, be sure to get at least one of each sex if you want them to produce berries. You may even want to plant several female plants, and only one male, for the most fruits. It is difficult to ascertain the sex of plants at any time except when they are in bloom. Only at that time can you readily identify the pistil on the flowers of the female plants, and stamens on the males. At other times you must rely on the word of the nurseryman. You may use asexual techniques to propagate plants of which the sex is known.

In some flowers, both the male and female parts can be readily identified. In a lily and similar flowers, for example, the female pistil, a single green object, shows plainly in the center of the bloom. It is hollow and leads to the ovary. It is surrounded by male organs, a cluster of shorter stems called stamens, which are covered with pollen. Wind, or insects, mostly bees, carry the pollen from flower to flower. After a grain of pollen touches the stigma or top of the pistil, it travels down its center tube to the ovum. Fertilization takes place, and the ovule eventually develops into a seed.

One grain of pollen must reach the ovum for each seed a plant produces. A peach, for instance, needs only a single grain for the flower to be fertilized and a fruit produced. An apple needs ten, and a great many more than that are needed for a thistle or dandelion bloom to be pollinated. Although the likelihood seems slim that so much pollination could take place, pollen is so small and plants produce so much of it, that even if only a tiny percentage finds its target it is enough.

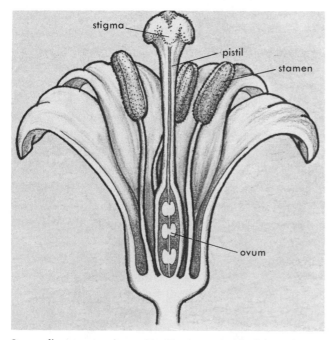

Some flowers, such as this lily, have both the male stamens and the female pistil and are pollinated by wind or insects.

Partners Needed

Some plants with perfect flowers need a partner to produce seeds, because Nature has ordained that they should not fertilize themselves. The pistil of most apple blossoms, for example, will not accept a grain of pollen from the stamens in the same bloom, or even from another bloom on the same tree. Another apple tree nearby must supply the pollen, or there will be no fruit. Furthermore, the two varieties must be different. If they are the same variety, they both were probably grafted from the same original tree, so the pistil would react as if there were only one tree present.

To make it even more complicated, the varieties shouldn't be too closely related. A McIntosh is reluctant to accept pollen from a Cortland, because the original Cortland was a descendant of the McIntosh.

Most fruit and berry-producing trees and bushes are pollinated by bees, as are most garden and wild flowers, flowering trees, shrubs, and many vegetables. Plants like fruit trees, which need cross-pollination, must be within bee-flying distance of each other. Although bees often fly for miles, five hundred feet is considered a more realistic distance for good insect pollination.

Wind Pollination

Grasses, grains, and most forest trees, on the other hand, are pollinated by the wind. The lightweight pollen of some plants can float for miles, as hay fever sufferers well know. The "sulfur showers" in pine forests, and yellow film sometimes seen on woodland lakes are further evidence of the abundance of pollen. Nevertheless, nut trees should be within 100 feet of each other to ensure pollination, and grains such as corn, must be within several yards. Corn and other grains should be planted in blocks rather than long single rows, so the wind can do an effective job.

EXPERIMENTS WITH SEEDS

New varieties of plants may be chance seedlings discovered by some observant horticulturist, or they may be the result of years of scientific breeding, either by amateur or professional plantsmen. The development of a new, superior variety of flower, fruit, vegetable, shrub, shade tree, or vegetable is as satisfying as any of life's great accomplishments, and even if you never win fame as a second Luther Burbank, plant breeding is a rewarding hobby.

To find a superior seedling is mostly luck, though they probably exist all about us. Although we may plant the seed of a named variety of rose, apple, or azalea with the greatest of care, the chances of growing anything worthwhile from it are very slim. The fact that there is a chance, however, is what keeps amateur horticulturists so dedicated to their hobby.

Scientific breeding, or hybridization, although it, too, involves an element of luck, is less a matter of accident. As with all breeding, the better the parents, the greater the chance for an outstanding offspring. If you were to plant the seed of an Imperial Silver lily, you'd know you had an outstanding mother, but the paternity would be in question. If you were to hand-pollinate the flower with the pollen from an equally outstanding daddy lily, however, your chances of its producing a superior offspring would be greatly improved.

Not only can two standardized strains of plants such as corn or wheat be cross-pollinated, but two hybrids can be crossed as well, such as a Reliance peach with an Elberta peach.

Contrary to the tales of some science fiction writers, however, plants must be closely related to cross-pollinate. The chances of crossing a kumquat with a zucchini are very remote, though the possible results might be interesting.

Every seed carries the genes of thousands of ancestors, and the seedling may pick up traits from any of them. Breeders must also keep Mendel's law of heredity in mind, and not be discouraged if the results are poor in the first generation. They know that it is often worthwhile to save seed from those seedlings, because often the outstanding plants they are seeking appear in the second or third generations.

How to Cross-Pollinate

(1) So the bees won't beat you to it, cover with a paper bag the blossom buds on the tree or plant you intend to pollinate, a few days *before* the blooms open.

(2) If the flower is bisexual, as most flowers are, carefully remove the stamens as soon as it opens. Use tweezers or tiny clippers, and be careful not to get even a drop of pollen on the green stigma, the top of the pistil.

(3) Go to the partner, the "male" member of this encouter. Brush the pollen from its stamens into a cup, using an artist's paint brush.

(4) Return to the first bloom. Dust some of the pollen from the cup onto the stigma. Cover the flower once again with the bag, so the pollen you put on will have a chance to enter the pistil before an unwanted grain comes along. Once pollinated, it will not accept any more.

(5) Mark the plant, limb, or twig carefully so you can identify it later, after the seed has ripened. If it is on a tree, you may want to pull off all the nearby flowers so there will be less chance of picking the wrong seed later.

(6) Remove the bag from the flower after a few days, and let the bloom fade and fall apart naturally.

(7) Keep accurate records of varieties and dates. If you come up with a real success, someone will want to know all about the parents.

Hybridizing Corn

With some plants, you don't need to go to all this trouble. If you choose to hybridize corn, simply plant alternate rows of two different var-

To cross-pollinate, cover with a paper bag the buds you plan to pollinate.

When blossom opens, remove the stamens, using tweezers.

Go to the "partner" plant and remove some of the pollen.

Return to the first blossom and dust some of the pollen onto the stigma.

Cover the blossom with bag again. This keeps out any unwanted pollen.

Remove bag after a few days, and let the blossom fade and fall apart.

ieties. Then pick off *all* the tassels, the male pollen producers, as soon as they appear on one variety. They are then unable to pollinate themselves, so pollen from the other variety, with the tassels left intact, will blow onto the silks of the detasseled stalks. The resulting corn on these detasseled stalks will be a hybrid of the two varieties. The row of corn with tassels will produce corn also, but it will be a mixture of hybridized and unhybridized corn—fine to use for food or fodder, but not as good for seed.

The fact that we individual experimenters cannot compete with the large research stations of big nurseries and universities should not be discouraging. Professional research teams are often interested only in developing plants of a commercial nature for big market growers. You may discover a berry, fruit, flower, or other plant that is ideal for home gardeners, and there is a large and often forgotten market for these plants, as well.

Experimenters have to keep things in perspective, too. Sometimes we come up with a good product and, naturally, get excited about it. A plant must be more than "good", however, to qualify as worthy of introduction. There are many good plants already on the market, so in some way our discovery or development must be truly superior to all the others. For more information about introducing new varieties, contact the state horticulturist at your state extension office.

DIVISION

When I started dividing plants, I thought of the process as similar to a stock market split. I had one share of rhubarb stock, and after a bit of work with a spade and a knife, I had gained four or five. It added interest to the propagating business.

Like seeds, division has long been used as a method of increasing plants. Much of the plant life we enjoy today was brought from Europe by early settlers. The number of plants they could bring with them on the boats was limited, so these were carefully divided and passed around. Even the Indians were soon planting European imports.

In its fullest definition, plant division can include any asexual method of separating a part of a plant from its parent, such as taking cuttings and grafts. The term as used in this chapter, however, is limited to the separation of new plantlets, suckers, offshoots, bulbs, scales, or tubers from an already established plant, or to the splitting of a clump of woody or herbaceous plants.

Doesn't Harm Plants

Some people who have never divided plants have qualms about doing it, feeling that taking away a piece of the original plant is in some way rather brutal and harmful to the plant. On the contrary, division is often necessary for certain plants, to keep them healthy and vigorous. It is also a simple way to get new plants. If you are hesitant to begin, start by dividing a potted chrysanthemum, or find a clump of perennials in your garden that looks easy to separate. As your

confidence grows, you will be ready to take on flowering shrubs.

It is easy to divide plants that form branches coming from below the ground, such as most herbaceous perennials, and many deciduous shrubs, bush fruits, vines, ground covers, and houseplants. It is much more difficult, however, to divide deciduous trees, many tropical plants, most coniferous evergreens, and shrubs such as azaleas, whose lower branches form mostly above the ground.

Study Growth Habits

To divide a plant successfully, you must examine its growth habits. Certain woody shrubs and herbaceous perennials, including mock orange and shasta daisies, expand the size of their clumps by growing numerous new plants all around them. These are easy to divide because the outer plants already have roots.

Plants with fleshy roots also increase in a manner that makes division possible. Lilies, daffodils, and similar plants can be periodically lifted from the soil, and their roots split apart and replanted.

Many plants increase by sending out suckers from their roots, and they are divided by digging up the sucker plants. Red raspberries and blackberries, for example, are easy to propagate in this way. Poplars, willows, black locust, wild cherry, and many other plants often send up new trees from their roots. Sucker plants sometimes sprout a long distance from the parent plant.

Others, including tropical plants such as pineapple and banana, produce sucker-type offshoots or offsets close to their trunk, which may be removed to start new plants. Offsets are sometimes tricky to start successfully because it is difficult to get a good root on each one when you cut it from the parent.

Many vines, ground covers, and similar plants, both herbaceous and woody, creep in all direc-tions, root, and create new plants in a process called natural layering. Wild grapes, kudzu, water hyacinth, and myrtle, for example, often spread over a large area in this manner. They can become either a useful covering for an unsightly eroded spot, or real pests. It is easy to dig up and divide the newly formed plants.

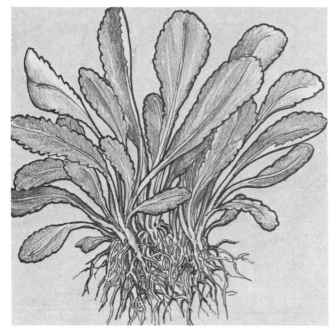

Clumps of plants are often the easiest to divide because the outer plants that you take off have roots.

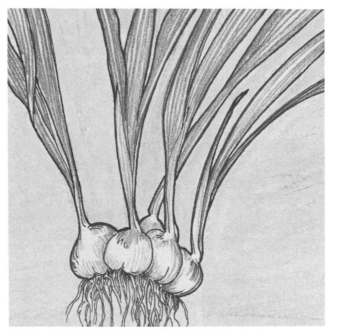

Fleshy rooted plants such as lilies and daffodils can be lifted from the ground and their roots split apart.

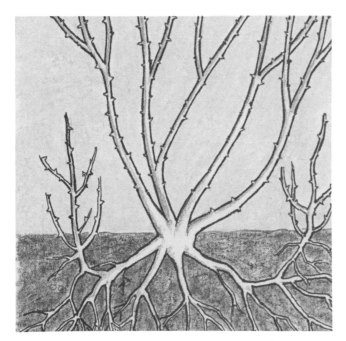

Plants such as raspberries send out sucker plants. Propagation is simple: just dig up the sucker plants.

Many wild plants such as cobweb houseleek send out creeping plants that take root. Dig up the youngest.

DIVIDING SUCKERING PLANTS

Digging up suckers is one of the easiest methods of propagating any plant that produces them, such as the common lilac, raspberries, and poplars. The parent plant is disturbed very little, and the transplanting is likely to be successful, since suckers are usually well rooted. In case the roots are not well developed, however, it is better to dig them in the spring, especially in the north, so the new plant will have a chance to become well established before winter.

Use a sharp spade or trowel to dig up a sucker plant, and try to get as much soil as possible to remain over the roots during the moving process. Cut the "umbilical cord" from the parent plant with hand pruners or a saw, depending upon the size of the root, rather than chopping it off with a spade. Less damage is done to the roots, so they heal over and start to grow faster.

Sucker plants usually have too much top growth for their new roots to support, so when you transplant them, cut back at least a third of the top to make a better balance. Even if there appears to be a good balance, pruning back the top will stimulate more rapid growth. Raspberries and blackberries need even more pruning and should be cut back nearly to the ground at transplanting time.

It is very important to keep the roots from drying out during the moving process. Set the sucker plants in well-prepared, fertile soil to help them off to a good start. Because a sucker plant is usually ill-equipped to compete with grass and weeds during its first months as an independent plant, a protective mulch of lawn clippings or shredded bark is helpful. Because of its fragile root system, daily watering and weekly feeding with a liquid fertilizer are beneficial during the early part of the summer. Stop all fertilizing by midsummer, so growth will harden before winter.

Dig up plant and the soil around it. Cut root with saw or hand pruners.

Keep roots moist at all times. Set plants in well-prepared fertile soil.

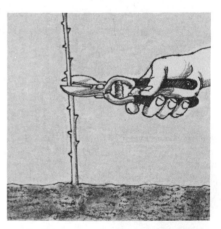

Cut back at least a third of the top. Roots can support this much top growth.

DIVIDING SHRUBS

Woody plants, including most flowering shrubs, should be split apart only when they are dormant. The best time to divide them in most areas is early spring. Vigorous small to medium-sized shrubs are best for dividing, but sometimes large old shrubs have smaller plants growing around them that can be dug up and moved.

Be sure the shrub you want to divide is not a grafted plant, or the small plants you split off may be of quite different variety than the main bush. Not many flowering shrubs are grafted, but French lilacs, hawthorns, and a few others occasionally are. If possible, choose your new plants from around the outside of the shrub, which is the younger, more vigorous section. Not only will the main plant be less disturbed, but, in addition, the central portion of the clump is usually woody and contains fewer feeder roots.

Check the Roots

First, before you attempt to cut the plant from the parent, dig around it carefully to make sure there are sufficient roots available. If there are, begin the operation.

Use a sharp spade for digging. If it is not too large, the entire bush can be dug up and cut apart. If you want, you can wash off the soil so you can better see where the roots are located. Use clippers to separate the smaller pieces and a pruning saw or hatchet to cut apart heavier sections. Make each division large enough so that it has at least one good root, and at least one sturdy branch. Don't let the roots dry out during the dividing and transplanting process, and plant them as soon as possible. Set them in the soil an inch deeper than they were growing originally, and cut back the tops about a third. Give them the same special care recommended for sucker plants.

If the new plant appears fragile, rather than plant it in the nursery or garden immediately, pot it up for a few weeks so you can give it extra care. It can then grow a huskier root system before it is planted out.

DIVIDING EVERGREENS

Most evergreens are propagated best by methods other than division, but a few prostrate evergreens, such as creeping junipers, can be divided, because they form roots on their outer branches. These can be dug, cut from the main plant, and replanted. Spreading yews occasionally root in the same way, and in northern hardwood forests, Canadian yews, sometimes called ground hemlock, often spread over large areas in this manner. Arborvitae, cypress, and hemlocks also form roots occasionally when a tree has fallen over and had some of its branches buried in soil or rotted leaves. Any of these can be separated easily from the parent plant. Replant as soon as possible after separating.

Mugho pines and other dwarf evergreens often look as if they could be dug and separated, but usually the branches have not actually formed any roots. The broadleaf evergreens, such as azaleas, rhododendron, holly, laurel, and similar plants, are not easy to start by dividing.

When you find an evergreen that looks as if it could be divided, dig carefully around it first, to make sure it has enough roots to survive the operation. If so, carefully cut it away and treat it as you would a hardwood shrub.

DIVIDING PERENNIALS

Perennials are divided more frequently than any other group of plants. Division is not only the common method of propagating perennials but, as any gardener with a perennial border will confirm, most clumps remain vigorous and healthy only if they are divided frequently. Asters, artemesia, chrysanthemums, geum, helianthus, hosta, and oenothera are some of the plants that must be divided at least once every two or three years if they are to thrive. Astilbe, campanula, daylilies, iris, lupines, pinks, phlox, platycodon, primrose, Oriental poppies, and similar plants usually need to be divided every three or four years.

Even slow-growing plants such as dictamnus (gas plant), old-fashioned bleeding heart, and peonies, need separating occasionally, usually every six to ten years.

Although it is possible for a skilled gardener to divide most perennials at any time, a beginning propagator will have better results by doing it at the time that is best for the plant.

Early Spring

This is the best period to divide all plants except those that bloom in early spring, and a few others, such as Oriental poppies and Madonna lilies. From the time the plant starts to show signs of life in early spring until the shoots are two or three inches tall is the ideal time to split most perennials.

Late Spring or Early Summer

Divide the plants that blossom in early spring, such as lungwort, doronicum, and primrose, immediately after their blooming period, so that the operation will not affect their blossoming the following year. The foliage of such plants as narcissus (daffodils) and other spring bulbs, should have died down and become yellow before the plants are dug and divided.

Summer

Oriental poppies, iris, peonies, and Madonna lilies respond best when they are separated after they have bloomed, and their foliage has discolored somewhat.

Fall

Most experienced northern gardeners frown on dividing plants in the fall, because it doesn't give the new tender plants time to get established before winter. In the South, fall is an excellent time to divide nearly everything except Oriental poppies and Madonna lilies.

There are occasions when you may want to divide a perennial clump at a time which is not ideal. Suppose someone wants a part of one of your special plants in midsummer when it is in full leaf and in bloom. If you can't talk the person into waiting, or if it is your Aunt Bessie from Schenectady whom you see only in July, take special precautions. Perform the operation on a cloudy day, if possible, or if not, at least in the cool evening. An hour before dividing it, soak the plant well, dig it with a large clump of soil, and wrap it in a wet newspaper surrounded by plastic. Warn the recipient to plant the division as soon as possible, water it frequently, and shade it with a pasteboard carton or plastic bucket for a few days.

Some perennials are far easier to divide than others. Clumps that are made of several individually rooted plants are very easy to break apart. Evening primrose and bee balm are good examples. Others, such as chrysanthemums, monkshood, anthemis, and daylilies, have loose crowns which can be split apart with little trouble. A few, including lupines and gas plants, have very tight crowns, and can be divided only with difficulty.

Separating Clumps

Perennial clumps may be separated by chipping off some of the plants around the outside with a spade, or they may be dug with a shovel or spading fork so the whole clump can be broken up. Try to get the plant out with the roots as intact as possible. After digging, you may wash

Avoid breaking roots when digging up parts of perennial clumps.

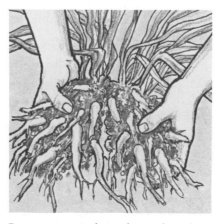

For strongest plants later, break the plant into good-sized chunks.

Cut back tops and keep plant at the same depth it was growing.

the soil away, if you wish, so you can better see what you are doing. I prefer to leave the soil around the roots, as much as possible, to help protect the fine hair roots.

Break the clump apart by hand, if the plants separate easily, or cut it into pieces with an old knife, scissors, or hand pruners. Be sure that you have some good sprouts and roots on each piece, and save only the best parts for your new plants. If part of the old crown is beginning to rot or deteriorate, discard it. Unless it is of utmost importance to get a maximum number of new plants, don't divide the plant into parts that are too small. Small plants are less likely to thrive. If you must start a frail plant, it is better to transplant it into a flat or small pot where you can give it special watering and feeding for a few weeks, before planting it in the ground. If the division is healthy and sturdy, place it wherever you want a new plant, at the same depth it was growing before. Keep it watered until it becomes well established.

DIVIDING BULBS & OTHER FLESHY-ROOTED PLANTS

Just as larvae develop into lovely butterflies, peculiar root and stem-like structures produce crocuses, lilies, dahlias, and other beautiful flowering plants. These corms, bulbs, rhizomes, tubers, and tuberous roots are primarily food storage organs for the plants which grow from them, as well as reproductive parts which can be divided or separated.

The following chart shows the type of structure of the most common plants.

TYPES OF FLESHY-ROOTED PLANTS

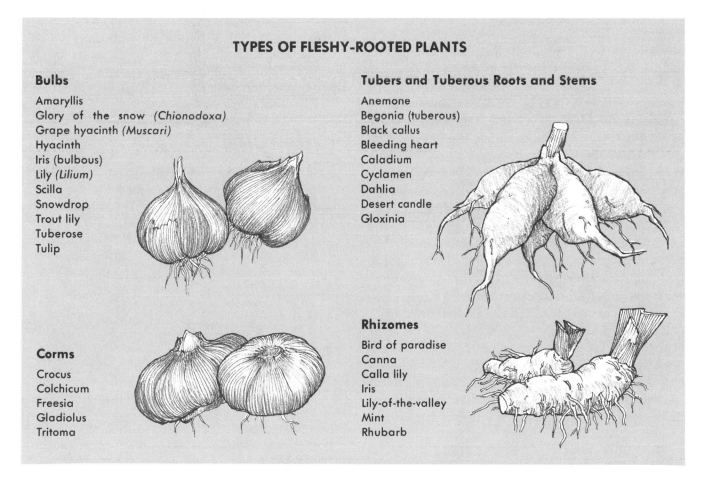

Bulbs

Amaryllis
Glory of the snow (Chionodoxa)
Grape hyacinth (Muscari)
Hyacinth
Iris (bulbous)
Lily (Lilium)
Scilla
Snowdrop
Trout lily
Tuberose
Tulip

Corms

Crocus
Colchicum
Freesia
Gladiolus
Tritoma

Tubers and Tuberous Roots and Stems

Anemone
Begonia (tuberous)
Black callus
Bleeding heart
Caladium
Cyclamen
Dahlia
Desert candle
Gloxinia

Rhizomes

Bird of paradise
Canna
Calla lily
Iris
Lily-of-the-valley
Mint
Rhubarb

THE BULB FAMILY

Bulbs are one of the most familiar fleshy-rooted plants. Everyone is aware of the bins of bulbs in the supermarkets each fall, and the resulting daffodils, tulips, and hyacinths are our symbols of spring, along with the potted Easter lilies.

Botonists describe bulbs as modified stems.

Bulb plants can be propagated in various ways. Although they may all be grown from seed, this is seldom done, except for the purpose of originating new varieties. It takes years, for one thing, and many plants do not come true. They are usually started asexually. The main bulb of most plants splits into two or more smaller bulbs, which can be separated and planted. Most bulb plants also grow numerous tiny bulblets around their main bulbs which may also be separated and planted.

If spring bulbs are to be divided, it should be done after they have finished blooming, and the foliage has died down. Lilies and other summer bloomers should be separated in the fall. Dig them with a spading fork, so you won't cut into the bulbs.

SPRING BULBS

Narcissuses (also called jonquils and daffodils) increase quite rapidly by themselves. Small bulblets form naturally around the original bulbs and, as the bulblets grow, the older bulbs gradually disappear. The clumps should be dug up and separated every three or four years. If they are not divided at least every five or six years, the plants will deteriorate because of overcrowding.

Like all spring bulbs, the best time to dig them is soon after the plants finish blooming. Wait until their foliage has deteriorated so the strength of the plant will have returned to the roots for storage. If they are planted correctly, the new bulbs should bloom the following year. Narcissus bulbs should be set pointed end up and about four inches deep unless they are being grown for producing more bulbs. Bulbs planted less deep multiply faster but bloom poorly.

Grape hyacinths increase naturally quite rapidly, and can be divided directly after blooming.

Large-scale production of tulips and hyacinths requires a favorable climate such as that found only in Holland and a few other places in Europe and America. Because there is always a wide variety of excellent bulbs available, and they are not easy to start, home gardeners seldom bother to propagate them.

The bulbs should be dug after they die down, following blooming, stored in a cool, moderately dry place, and replanted just before the ground begins to freeze slightly in the fall. Tulips form new bulblets around the old bulb, but unless the climate is right, they fail to develop well.

Hyacinths, too, should also be dug up following blooming, and treated the same as tulips. Where climatic conditions are right, they also increase by bulblets which can be separated and

DEPTH TO PLANT BULBS

If planted at the proper depth and in soil that provides sufficient food, a collection of bulbs will provide blossoms from early spring to late fall.

The crocus and grape hyacinth provide early blooms, even before the final frost, lilies and many others will blossom in the summer, and late fall flowers will include colchicum and autumn crocus.

Place bulbs with the pointed end up. That end should be buried this deep:

Bulb	Depth (inches)
Crocus	1
Snowdrop	1
Tulip	1½
Narcissus	2
Hyacinth	4
Lily	4

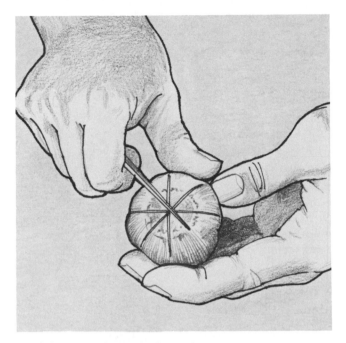

Notch hyacinth bulbs before planting to increase production of bulblets.

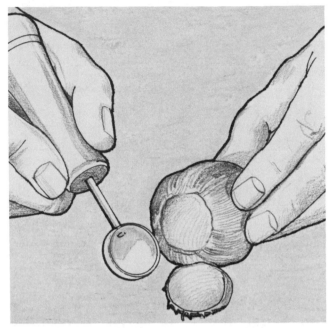

Or scoop out a portion of the base before planting it.

replanted in beds or flats until they are large enough to plant out in the field. Unlike tulips, their supply of bulblets can be increased by wounding the bulb. Before planting in the fall, cut a few notches in a pie shape and about an eighth of an inch deep in the bottom of a large hyacinth bulb. Notching allows more surface

where bulblets can form. They can be removed and planted when the bulb is dug the following year. Some propagators, instead of notching the bulb, scoop out a portion of the base of the bulb instead. This increases bulb production also, though not as much as the notch method. The new bulbs tend to be larger, however.

LILIES

Some of the more common varieties of lilies, such as the Regal, are sometimes propagated by seed, but most members of the lily family are increased by different kinds of division.

Dividing the Bulb

Lily bulbs are scaly, and resemble a globe artichoke. Every year or so, the large bulb splits into two or more bulbs and the bulbs of some varieties develop new bulbs which grow inside the old ones, pushing them apart. These bulbs can be easily separated in either spring or fall, and each quickly grows into a large plant which usually blooms the following year. This is an excellent way to propagate lilies if only a few are needed.

Planting Bulblets

In addition to splitting periodically, bulbs also multiply by growing numerous small bulbs, called bulblets, around the larger ones. Bulblets will increase in size somewhat if left where they are, but to avoid their being stunted by overcrowding, separate them carefully from the parent bulb in early fall. Replant them at twice their own depth in a bed or flat of sifted rich, sandy soil, and grow them for a year or two to increase their size. They can then be planted permanently at the proper depth recommended for that variety of lily. This final move should be made in the fall.

Peel off scales in spring or fall. Plant them, points up, ½-inch deep.

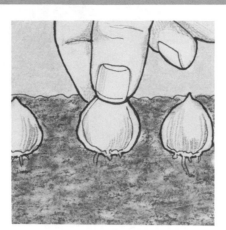

A sprout will first appear, then a small bulb forms around scale.

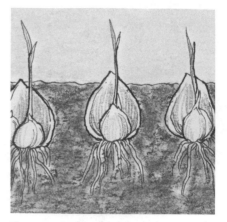

The new bulb can be dug up and planted a year after starting.

Planting the Scales

The scales that make up the bulb can be removed to increase your lily supply. They can be taken from bulbs you buy, or from your own. If you don't want to disturb a favorite bulb, you can even dig down into the ground around it and pick off a few scales.

Scaling can be done in either spring or fall. Peel several outside scales from the main bulb, and plant them points upward, a half-inch deep in a flat or in a well prepared bed outside. After a time, a grass-like sprout starts to grow, and a small bulb forms at the base of the scale. The scale will gradually disappear, and a year or so later, the new bulb can be transplanted.

Planting Bulbils

Another way to propagate lilies is by bulbils. Not every variety produces these, but the Madonna, tiger, and several others do. Small bulbs form in the leaf axils along the main stem of the plant. They fall off in late summer, and if they land in a place favorable for growing, they will grow into large plants. If you want to propagate these lilies, remove the bulbils as soon as they are ripe enough to separate from the stem easily, and plant them a half-inch deep, in the same way as bulblets and scales. They, too, will grow into bulbs.

With certain varieties it is possible to increase bulbil production by removing the flower

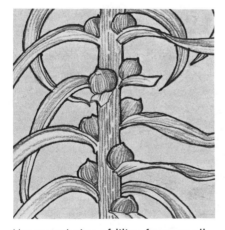

Many varieties of lilies form small bulbs in their leaf axils.

Separate them from the plant, then plant them a half-inch deep.

Or bury the stalk an inch deep. Little bulbils will grow along it.

bud before it opens. This procedure sends extra energy into the stem to make bulbils.

Some growers pull the stalk from the bulb directly after the flowers fade (it doesn't seem to hurt it any), and bury it an inch deep lengthwise. Little bulbils grow along the stalk and produce new plants right where it is buried.

Planting bulbils is an excellent way to propagate rapidly the plants that produce them. The tiger lily increases so rapidly by bulbils that only a short time after their introduction, the lilies had become an escaped "wild" flower, and are now often found growing along country roads far from civilization.

CORMS

Corms perform the same function as bulbs but are structurally quite different. They are shorter and broader than a bulb, and are enclosed in a thin dry membrane that protects them from drying out. As a corm increases in size, it develops more sprouts, and as long as there is a sprout on each portion, a large corm can be cut into two or more pieces, each of which is capable of growing into new plants.

They can also be increased by planting the cormels, the small corms that form around the larger one. Most of these are too tiny to bloom the first year, but you may plant them three-quarters of an inch deep in beds or flats where

they will grow large enough to bloom the following year.

Gladioli are a good example of the plants that grow from corms. In the fall, after the corms have been dug, dust them with an insecticide and store them for the winter. In the spring, they can be divided before you plant them. Cut apart the biggest corms with a knife, making sure there is a sprout on each piece, and separate the small cormels from the large corms and from each other. Plant them sprout-side up. The medium-sized corms will probably produce blooms the first year; the smaller ones should be grown for a year as described earlier.

Cut corms into two or more pieces, each with a sprout on it. When planted, each piece will grow into a new plant.

Plant these sprout side up. The larger pieces will blossom the first year; the smaller pieces will take another year.

TUBERS AND TUBEROUS-ROOTED PLANTS

Tubers are quite different in appearance from bulbs and corms. Potatoes are one of the most common tubers, but Jerusalem artichokes, dahlias, and old-fashioned bleeding hearts all have tubers or tuberous roots. When dormant, this class of plants has "eyes" which are either scattered over the plant as on potatoes and artichokes, or situated at one end, as on dahlias. On other plants, such as tuberous-rooted begonias, the sprout is at the top.

Clumps of dahlia roots may be separated into small pieces in the spring after winter storage in a cool, dry spot. Since it is important to get a sprout or eye on each piece, it is better to cut them apart with a knife rather than attempt to break them by hand. If you plant tubers that are too small, they may not bloom that year.

Although tuberous begonias and gloxinias are started from seed when large quantities are wanted, small numbers can be obtained by cutting apart and planting the mature tubers. Wait until the sprouts show plainly before performing the surgery.

It is often possible to get two or three new plants from each large root. Cut it apart as if you were slicing a pie crosswise.

RHIZOMES

Certain garden perennials, weeds, and food plants that we usually think of as having ordinary roots actually have rhizomes. A rhizome is a stem that grows laterally at the soil surface, or slightly above or below it. The iris family has them, as do peonies, rhubarb, mint, quackgrass, and some ferns.

The method of growth of these plants varies somewhat. Some have crowns, such as rhubarb and asparagus, while others, such as mint and quackgrass, make rapid growth, and create new plants in all directions. Crown rhizomes are more difficult to propagate than the running kinds, but the crowns can be cut apart if the operation is done carefully, making sure there are eyes (sprouts) on each division.

The new plants should be set at the same depth they grew originally. With some, like iris, this means planting nearly at ground level. See the herbaceous section in part two in this book for specific directions for the plant variety you wish to propagate.

DIVIDING FOOD PLANTS

Many of the perennial herbs like mints and French tarragon, and food plants such as rhubarb, Jerusalem artichokes, and horseradish are easily propagated by division. Some, if they are not split up occasionally, become rootbound and unproductive. Although most can be divided in late summer after the tops have stopped growing and have either changed color or died down, early spring before the foliage shows is probably the best time. The ground is moist then, and conditions are right for the new plants to begin to grow rapidly. The rugged plants can be dug up entirely and cut into pieces, each with a sprout or eye; or you may dig into the clump with a sharp shovel and chip off pieces for planting. Horseradish, mint, and Jerusalem artichokes can be treated quite roughly and still survive. Although chipping off parts of the clump will not produce as many plants as digging and cutting apart the whole plant, the original root is left relatively undisturbed, and will, if you're careful, produce the same year.

Asparagus and rhubarb must be cut apart carefully to be sure of getting the best plants, and it is better not to make the divisions too small. Plant them to the recommended depth, usually with the crowns three inches or more below the surface of the soil, and water them thoroughly. Both plants thrive with an abundance of organic fertilizer and it is difficult to give them too much manure.

Any stock plants you raise especially for dividing thrive on manure too, so if you want a lot of new rhubarb, asparagus, and horseradish plants, I recommend a heavy feeding for the parent plants as well.

Woody food plants may also be divided. The bush fruits such as currants, gooseberries, elderberries, serviceberries, and bush cherries usually produce numerous small plants around the main clump which can be split off and planted. If a great many plants are needed, and the whole bush is dispensable, it may be dug up, and clipped or sawed into several pieces.

Wild blueberries sometimes send up suckers or offshoots, which may be dug and transplanted. As a rule, very few suckers appear on the cultivated highbush varieties, and they are difficult to separate successfully, because most have a scant root system. Many good blueberry plants have been wrecked when amateur gardeners tried to cut them apart to make more plants. Highbush blueberries are usually propagated, instead, by softwood cuttings.

DIVIDING HOUSEPLANTS

Although some houseplants can be easily divided, many cannot, and any attempt to cut them apart might be fatal. Vines, plants with woody stems, and those that spring from a single stem cannot be divided. Usually divisions can be made of those plants that make a thick, spreading growth near the soil surface such as fibrous-rooted begonias, many ferns, clivia, and chrysanthemums. Those that send up small plantlets around the main plant can also be divided.

Divide Plants in Spring

The best time to divide houseplants is usually in the spring, when they are beginning to grow. For a few days, before you perform the operation, give a plant less water than usual to help "harden" the top growth.

When you are ready to divide the plant, tip it gently out of the pot, and cut the various sections apart with a sharp, sterile knife. Clippers may be used, but they tend to crush the fragile stems. Make sure there is a good root and top on each section. As with perennials, for obtaining the most plants, you may want to wash the soil off the roots so you can better see where the dividable sections are located. Unless the maximum possible number of plants is your goal, it is better to leave the soil on the roots and divide the plant into no more than two or three strong units. Work quickly, don't let the plants dry out, repot the new divisions in good, fresh soil, and water them as quickly as possible.

Some tropical plants that are grown in pots send up offshoots, tiny complete plants, that can be separated and grown as new plants. Shake or wash away the soil to make sure the offshoots have good roots before attempting to cut them off, since some of them form roots rather slowly.

LAYERING

Layering is a process whereby roots are induced to grow on an outside branch or stem of a plant while it is still attached to the parent. After the roots are formed, the branch is then separated to form a new, autonomous plant.

Soon after I was married, I discovered that a block of my carefully layered clematis vines had been dug up and put back in place on the trellis. Suspecting it was much too neat a job for the cat or a wild rabbit, I accosted my wife who had only recently become acquainted with gardening.

"I thought someone had messed up those nice plants by burying them," she explained, not noticing the nice little roots all along the branches, now drying out in the summer sun. My voice was somewhat strained as I introduced the process of layering to her, under less than ideal circumstances.

Layer Themselves

Throughout nature there are examples everywhere of plants that layer themselves. Strawberries send out little runners that form new plants which root wherever they touch the soil. In some areas, acres of wild grape vines trail over the ground, rooting and creating more plants. Fallen trees, or branches bent down by snow, sometimes grow roots as their limbs become covered with soil or leaves. Gardeners have long used this natural process for the purpose of propagating new plants.

Layering is a useful method for gardeners to start limited numbers of plants without expensive equipment and with little attention. It is also often used by large commercial nurseries for starting varieties for which the demand is limited.

We propagate flowering quince, gooseberries, currants, cotoneasters, black raspberries, dwarf viburnums, daphnes, Japanese tree lilacs, shrub roses, and many others by layering. Certain plants, however, seem to resist forming roots on a layered branch, so they must be started by seeds, cuttings, or in other ways. Mountain ash, most conifers, broadleaf evergreens, and high-bush blueberries are in the category of those that do not root easily by layering.

Time Varies

The time it takes for a layered branch to root varies greatly. Grapes, currants, mock orange, and ivies root quickly. If layered in early spring, they will usually form good roots before the end of the growing season. Fruit trees, maples, and French lilacs take a long time, sometimes several years to form roots. Trying to discover the reasons why some plants root so readily and others not at all is frustrating.

To layer a bush or tree, it must have branches close enough to the ground to be bent down and have their midportion buried in the soil. If the plant you wish to layer has no such branches, it may be possible to get some to grow. Cutting back the tops in early spring, when the plant is still dormant, often stimulates the growth of lower branches. By the following year, some of them should be long enough to layer.

If you prune back the top, keep in mind that no tree or shrub should be overpruned, and never cut off more than the amount recommended for that particular plant. This means usually cutting off no more than one-third of a young plant and considerably less from an older one.

Start in Spring

Layering can be done anytime, but unless it is done before the early summer growing season starts, roots are not likely to form until the following summer. First, bend over the branch and bury its middle portion about three to five inches deep in loose, fertile soil. Remove any leaves from the part that is to be buried. Branches no larger than a quarter inch in diameter root best, but it is possible to get much larger ones to root. To encourage the fastest rooting, cut away a piece of bark from the underside of the part of the branch that will be buried in the soil. The cut should be about twice the size of the diameter of

the stem. Dust the wound with a root-promoting product such as Rootone. This wounding, as it is called, plus the chemical treatment, often encourages even hard-to-root plants to layer successfully.

Prepare the soil well where the branch is to be buried. The more thoroughly it is loosened initially, the better the new roots will grow. Mix in some sand, if the soil is heavy, and add dried manure or compost to help feed them.

I have found that composted leaf mold or natural leaf mold from a hardwood forest is an ideal rooting medium. Rather than burying the branch in the soil beneath, I remove a portion of the soil and replace it with the compost. After rooting has started, apply liquid fertilizer to stimulate fast growth.

To layer a plant, cut off leaves from a branch that is near the ground.

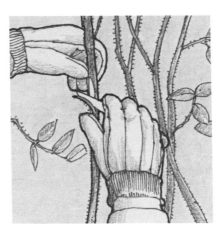

Remove a piece of bark from the underside of the branch.

Dust the wound with a root-promoting substance such as Rootone.

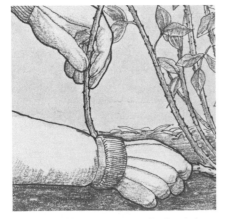

The tip will grow erect if placed against the side of a shallow hole.

To get the tip to grow erect, place it against a stake.

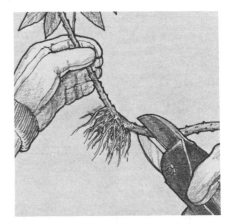

When roots have formed a mass, cut the layered section from the plant.

Keep Tip Erect

A better-shaped plant will result if, when burying the branch, you force the tip of it to stand up as straight as possible. This can be accomplished by digging the hole with the far side at a vertical angle. Put a small stake beside the tip of the layer, to keep the new plant upright while it is rooting and growing.

Place a flat rock or brick on the soil over the buried portion of the branch. It not only prevents the branch from popping up, but also helps keep the soil beneath it moist. In dry seasons, water the soil around the layer occasionally.

Rooting occurs when certain organic materials are interrupted in their passage from the plant to the tip of the branch. They accumulate at the point where the rooting takes place. The conditions that help bring this about are adequate moisture, well-aerated soil, warm temperatures, and the absence of light.

To Encourage Rooting

One method used to encourage rooting on fruit trees and other plants that form roots with some reluctance, is to twist a small piece of fairly fine wire tightly, but not so tight as to cut into the bark, around the branch at a spot somewhere between the part that is layered and the main plant. As the limb increases in size, the nutrients from the plant are gradually cut off from the layered section of the branch by the constriction of the wire, and the branch is thereby encouraged to develop its own roots in order to survive.

Check for Results

From time to time check to see if rooting has taken place by carefully pulling the soil away from the layered area. Firm it back into place promptly, before it, or the layered area of the branch, has a chance to dry out. As soon as a good mass of roots has developed, cut the layered branch from the main plant with a pair of hand pruners. Make the first cut just above the soil surface, and then cut back the remaining part of the branch to a leaf or the main trunk, so no dead stub will be left.

Layering can also be done in pots instead of in the ground, either in a greenhouse or outdoors. Fill a plastic or clay pot nearly to the top with fertile, sandy soil mixed with some composted leaves or peat moss. Place the pot beneath the part of the branch to be layered, and proceed

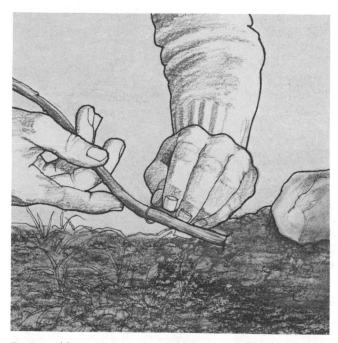

To speed layering, wrap wire tightly around branch. This will cut off nutrients, encourage root development. Wire must be between plant and layered area.

To save one step in the layering process, it's possible to layer plants directly into a container. This is often done with strawberries, which are sold that same year.

with the process as if there were no pot. Outdoors the pot can be sunk in the ground to save watering. Otherwise it should be watered daily.

Plants layered in this manner can be removed from the parent plant immediately after they are well rooted, and then be planted out, even in the summer, without waiting for the tree to become dormant. Strawberries, grapes, ornamental vines, black raspberries, and ground covers are often rooted in pots in plant nurseries, so they can be sold the same year.

Some gardeners prefer layering to division, because they feel that it does the plant less damage. The low-growing branches that are layered are often ones that should be pruned away anyhow, so nothing of value has been lost. Even when layering is done annually, the process doesn't seem to hurt a healthy plant.

TIP LAYERS

In this form of layering, the tip of the branch is buried, rather than the middle portion, and it is done mostly on plants belonging to the *Rubus* family. Black and purple raspberries, vine blackberries, dewberries, boysenberries, and loganberries tip layer naturally, by bending over their long, limp canes and rooting where they touch the ground. The tip also grows a new top, forming a complete new berry plant.

By the following spring, this plant can be cut off and transplanted.

If you want to propagate plants faster than is happening in the natural process, you can "tip" additional canes, in mid-summer, by bending over and burying their top ends five or six inches deep in soil that has been well spaded and enriched. Propagate only from disease-free plants, and be sure to choose canes that grew that season, rather than those that are currently bearing fruit. Firm the soil around the top, or tie it to a stake, so the cane won't snap out of the ground in a high wind.

In early fall, after the tip has rooted and started to grow, cut the new plant from the cane, just at the surface of the soil. It will then develop as an independent plant.

Can Be Potted

If you want a potted plant, stick the tip of the cane into a six- or eight-inch pot that has been sunk into the ground and filled with rich, light soil. The resulting rooted plant may be set in the garden without being pruned back, or it makes a nice item to sell.

COMPOUND LAYERS

Vine-type plants and ground covers are sometimes rooted in many places along their flexible branches instead of only in one spot. This makes possible the propagation of many plants in a small area, with little fuss. Clematis, grapes, ivy woodbine, honeysuckle vines, wisteria, rambler roses, myrtle, and Dutchman's pipe, as well as most other vines and ground covers can be started in this manner. It is called compound or serpentine layering.

Layer the vine in early spring in the same manner as you would a branch, by burying sections a foot or so apart in soil that has been loosened for easier rooting. Each buried section of the branch should be adjacent to a bud or shoot that is not buried. The shoots will develop into the new plants. Cover the buried parts with flat rocks or an organic mulch which will speed rooting by keeping the soil moist. By the end of the growing season, roots should have formed at each buried section. Compound rooting may also be done in pots.

The vine can be dug up the following spring, and the rooted sections cut apart. The new plants

Compound layering provides many plants on one branch. Bury sections a foot or so apart, leaving shoots that are not buried and that will grow.

Cut plants apart when each has roots. Replant them in the spring.

can then be trimmed to a neat shape, and either planted, or potted and sold.

It is also possible to simply lay along the ground some of the more durable vines, such as honeysuckle and Virginia creeper, and place a shovelful of soil over them at spots between each clump of leaves. Cover each pile with a flat rock or bricks. As soon as a good root system has formed in each pile of soil, the vines can be cut apart. In spring they can be dug and replanted. I have used this method successfully with easy-to-root ground covers, such as myrtle.

STOOL LAYERING

Stooling is similar to simple layering, but the entire shrub is layered by being buried where it is already growing, rather than burying only part of one branch. Like layering, it is a good method for home nurseries because it requires no greenhouse or mist system, and very little attention.

It is possible to start a greater number of new plants from each mature plant by stooling than by layering. The plants produced are more uniform in size, too, than those started by the usual layering method.

The stooling technique is used by commercial nurseries for growing certain shrubs and small fruits that they sell in small quantities, and for starting rootstocks for propagating fruit trees. Large numbers of dwarf trees, especially the popular Mallings, are grown in this way, and are budded or grafted before being sold.

The stooling method works best with young, bushy, vigorous, deciduous plants, not more than three or four feet high, although fruit tree understocks taller than that are sometimes stooled. If trees such as apples and plums are to be stooled, they should be pruned back heavily the previous year so they are growing in a bush form, rather than with a single trunk. This will make them grow even more new sprouts, when cut back severely for the stooling.

I have had good luck stooling shrub roses, bridal wreath and other spireas, currants, gooseberries, viburnums, potentilla, quince, Japanese lilacs, red leaf plums, and apple understocks for grafting. I have had only fair success, however, in stooling some of the named varieties of apples, plums, cherries, red leaf maples, bayberry, cotoneaster, and daphne. These took at least two years to root well.

To stool a plant, in late winter or very early spring cut it back to about two inches above the ground. New shoots will sprout from the cut ends. If a lot of these start, just let them grow. If only a few sprout, pinch off the new bud at the end of each one, so they will stop growing temporarily, and the plant will send up additional shoots.

When the new shoots are about six inches tall,

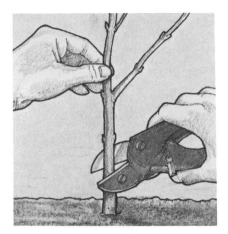

To stool a plant, in late winter or early spring, cut branches back to about two inches above the ground.

Shoots will sprout from these cut ends. If there are only a few, pinch off bud at the end of each one.

When sprouts are six inches tall, mound rich, sifted soil around them so that they are almost buried.

As the shoots continue to grow, add more soil, creating a mound that is six inches to a foot in height.

To hold in moisture and prevent erosion, cover the stool with a mulch. Use shredded bark or lawn cuttings.

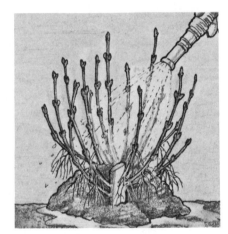

When plants have their own roots, wash soil away from them, using a hose or several buckets of water.

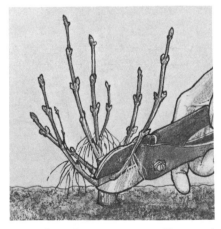

Using hand pruners, cut off rooted plants just above the stems of the original plants. Leave no stubs.

Transplant quickly so that roots don't dry out. These young plants can be placed in pots or the garden.

To save the stool plant, fertilize it with a light sprinkling of 5-10-10 or work dried manure in soil around it.

gently pile light, rich, sifted soil around them so they are almost, but not quite buried. Pack the soil gently between each shoot. As the shoots grow, add more soil until the mound is between six inches and a foot high, depending on the height of the plant.

Cover the completed stool with a mulch of shredded bark or lawn clippings. The mulch helps hold the moisture in a dry year, and helps prevent the soil from washing away from the plant during a hard rain.

Check the stools occasionally to make sure they are not drying out, and water them when necessary. Don't overfeed them, because there is already a great amount of root area for a small amount of leaf surface, but if growth is poor, give them an occasional dose of liquid fertilizer in mid-summer.

On many kinds of shrubs, roots start to form on the buried ends of the new sprouts in a few weeks. As with layering, you can dig away some of the soil by mid-summer, if you do it carefully, to see what is happening. If all goes well, the roots will be heavy enough so the new plants can be separated the following spring.

If the plants take more than a year to root, you will probably need to add additional soil to replace any that has washed or blown away.

To harvest your crop of new plants, gently wash away the mound of soil with a hose, or buckets of water, and clip off the rooted plants just above the stems of the original plant, cutting close enough so no stubs are left. Transplant the new plants to pots, or set them into the ground.

Then fertilize the stool plant with a generous helping of dried manure or with a light sprinkling of 5-10-10. You can then either allow it to grow back into a bush or tree, or stool it to grow another crop of plants. Stool beds often produce new plants for twenty years or more without deteriorating if they are kept fertilized and well maintained.

Burying The Entire Plant

Similar to stooling is a process the English call "dropping." In this method, the entire plant is buried. Although it destroys the original plant, it is useful if you need a lot of plants at once, or you want a grafted tree to grow on its own roots. Small bushy plants such as heathers, heaths, dwarf rhododendrons, and andromeda are sometimes propagated this way.

If the plant isn't already nice and bushy, it should be cut back a year or more ahead of time, to encourage a lot of short branches. Then in early spring, dig up the plant carefully with its soil-rootball intact. Set it horizontally in a much deeper hole, large enough so that the top ends of its branches protrude only a few inches above the soil line. Fill in around the branches with rich, light, well-sifted soil. The buried ends of these branches will then grow roots.

The following spring, dig up the entire plant, wash off the soil, and "harvest" the crop of young plants, as in stooling.

A variation of this method is used to get grafted trees to grow on their own roots. Only one- or two-year-old grafted trees should be used

In dropping, cut back plant a year in advance to encourage new growth.

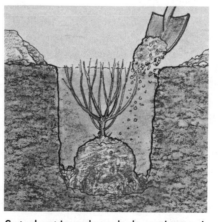
Set plant in a deep hole so that only its top protrudes.

Fill around branches with well-sifted soil. The buried ends will grow roots.

Partially bury the shoots when they are five inches tall.

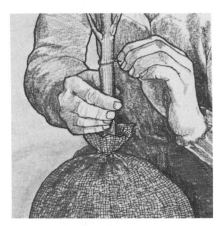
Another method is to twist a wire tightly around the trunk of a tree.

Tree is then placed in a trench, with roots covered, branches bare.

for "dropping" since older trees would be too large. Dig the tree in early spring, while it is still dormant, or use a newly purchased one. Twist a piece of wire tightly around the trunk just above the bulge where the tree was budded or grafted. As the tree grows, the wire will constrict sap flow and stimulate roots to grow on the buried branches. Dig a large trench and lay the tree horizontally in it. Place the tree as flat as possible and make sure it is entirely inside the trench, even if you have to prune off some of the branches on the lower and top sides. Bury the roots carefully, but leave the remainder of the hole unfilled and branches uncovered.

As the weather warms, the tree will send up little upright growing shoots from the laid out branches. When the new growth is about five inches tall, add enough soil to the trench to par-tially bury the new shoots. Use light, rich, sifted soil, so the new roots can form easily in it. Keep adding soil as the shoots grow, until the trench is filled to ground level. Water it whenever the weather is at all dry, and roots should eventually form at the base of each sprout.

Don't check too often to see if roots are forming. It may take two years or more to develop a good set of little trees with roots heavy enough so the separate plants can be safely cut apart and planted as individuals.

When you are sure that the root growth is sufficiently heavy, wait until the following spring, and exhume the entire plant. Wash off all the soil, and discard the original rooted part of the tree up as far as the wire. The remaining part should consist of several potential good trees that can be cut apart and planted.

AIR LAYERS

Some woody plants such as tall shrubs and flowering trees are difficult to propagate by layering because they have no branches that can be bent over to root in the ground. Often it is possible to root these high branches, however, by bringing the earth to them in a method called air layering. Although only one layer can conveniently be placed on each branch, on a large plant a great many branches can be air layered at once, so a lot of new plants can be started.

Originally, air layering was a method used by the Chinese, who placed a pack of wet soil around an upright growing limb or stem, wrapped it in cloth or burlap, and kept it moist until roots formed.

Western gardeners later improved the method by sawing a clay pot in half, from top to bottom, and reassembling it around the limb. The pot was filled with a mixture of soil and sphagnum moss and kept moist. With the development of plastic, air layering became much more simple, and easy for the beginner to try.

Air layering can be done at any time, but it is best to do it in early spring, either just before or just after the leaves come out. The best place to position an air layer is on the new growth of the upper branches of the tree and, of course, the leaves should be left on the branch because they aid in the rooting.

About ten inches from the end of the branch you intend to air-layer, scrape away a bit of bark to make a wound all the way around the branch and about a half-inch wide. Dust the wound with rooting powder, and wrap a handful of moist sphagnum moss around this section of the limb. Then wrap a small piece of black polyethylene around the moss, and seal it tightly along the edge and at the ends with plastic electrical tape or furnace duct tape. A collection of these little black bags look odd all over the tree, and the first year I tried it, my customers were suspicious that some new disease had struck our nursery.

Some trees and shrubs may form roots quite fast, but others may take a year or more. If you unwrap the bags to check, be sure to seal them back tightly. Sometimes it is possible to feel the roots inside, but don't squeeze too hard.

As soon as the roots are well formed, cut the branch from the tree, just below the bag, and unwrap it. If you have a mist house, take the branches off in mid-summer, pot them up, and put them under the mist until they become well

To air layer, scrape away a half-inch band of bark on new growth of upper branches of the tree.

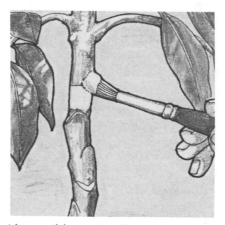

It's possible to stimulate growth of the new roots by dusting with a rooting compound, such as Rootone.

Moisten a handful of sphagnum moss, squeeze out moisture, then wrap the moss around branch.

Wrap a piece of black polyethylene plastic around the moss. Seal it tightly on all sides with tape.

When roots are well formed, cut off the branch just below the bag, then carefully remove plastic sheet.

Pot up the new plant, then cut back top growth so roots can support it. Store in cool area over the winter.

established new plants. If you have no mist system, wait until late fall to cut the rooted layers from the tree. Then pot them and store them in an unheated greenhouse, cold frame, or root cellar over the winter. The branch above the layer will probably have made a lot of growth while the roots were forming, and consequently there will be more top than the roots can support. Before storing, cut the top back by about half, so the plant will be in better balance and can start growing faster in the spring.

My first attempts at air layering produced disappointing results because I was not careful. My first mistake was having the moss too wet. All the water should be squeezed out, leaving the moss only slightly moist. The plastic wrap should be sealed very tightly, not only along the edge, as I had done it, but at both ends where it touches the branch, so the moss won't dry out and so rain cannot enter. In my impatience, I unwrapped the layers every few weeks to see how they were progressing, and my unwrapping let in even more air.

French lilacs, fruit trees, many shade trees, rhododendrons, and magnolias can be air-layered successfully, but the method is used mostly to increase potted plants that are difficult to start in other ways.

Rubber plants and numerous tropical plants are often air-layered in greenhouses. For most outdoor plants, there are faster and simpler methods, but it is an interesting hobby and a convenient way to get grafted trees to grow on their own roots.

CUTTINGS

Nearly everyone has had some experience with cuttings. Starting a new geranium or ivy from a "slip" is a familiar process, and most of us have heard the story of the farmer who stuck his willow cane into the ground one day and it grew into a landmark tree.

A cutting is part of a branch, root, or leaf that is taken from the parent for the purpose of starting a new plant. It is placed in some kind of medium and, if all goes well, in time it will form roots and grow into a new plant, a replica of the parent.

Cuttings are a useful propagation method for plants that produce few seeds or no seeds at all, and for those that are difficult to grow from seed. It is also advantageous to start from cuttings plants that are usually propagated by grafting, since a plant from a cutting has all the characteristics of its parent, but one started from a graft invariably acquires certain characteristics from the rootstock. With this method, a great many new plants can be started from a single large plant without disfiguring it.

Not everything roots as easily as willows and geraniums, of course, or there would be little need for nurseries. In theory, any plant can be grown from a cutting, but in real life, some can be downright stubborn about rooting. Rhododendron cuttings, for example, need a great deal of special care if they are to develop roots, and I've found that those from a mountain ash usually refuse to root at all, no matter what method I use.

There are many types of cuttings, and although plants often root by more than one method, most species respond best to one particular type.

MEDIA

Plants not only appear to reproduce best from one specific type of cutting but some also have a favorite medium.

A few form roots while submerged in water. Some houseplants, such as ivy, start well in tumblers or jars of water, and certain woody shrubs such as willows, rambler roses, oleander, and poplars can also be started in this way. The water should be kept at approximately 70 degrees F. and not be changed during the rooting process. Root production will be stimulated, however, if you remove the cuttings every few days and stir the water vigorously to mix air with it. My wife covers the container with a piece of foil and sticks the cuttings through the foil, so they will stand upright instead of leaning against the edge. Plants rooted in water should be removed and potted as soon as the roots become well formed, because they rot quickly if left too long.

Most cuttings root best in a well-drained material that keeps them moist and provides other conditions conducive to root formation. Although a good medium should hold moisture, it should not retain such great quantities that the cuttings will rot. It should be loose enough for fast root development and should not support harmful disease organisms.

Sharp, clean river sand has been used as a rooting medium for centuries, and is still favored for many plants, especially houseplants and evergreens. Peat moss, sphagnum, perlite, vermiculite, soil, composted leaf mold or bark, sawdust, and other materials are also used either alone or in various combinations, depending upon the requirements of the plant to be rooted.

Rooting Media

The most common rooting media are the following mixtures (by volume):

(1) 1/2 sand and 1/2 peat moss

(2) 1/2 sand and 1/2 vermiculite

(3) 1/2 perlite and 1/2 peat moss

(4) 1/2 perlite and 1/2 vermiculite (our favorite for most things)

(5) 1/3 sand, 1/3 perlite, and 1/3 either vermiculite or peat moss

Neither soil nor homemade compost should be used as a medium to start the cuttings of most plants, because they harbor disease. Cuttings that root quickly can be rooted satisfactorily in either, however, and we have started successfully such varied plants as willows, poplars, chrysanthemums, petunias, tomatoes, periwinkle, and peppers in a sandy-soil mix. Although it is possible to sterilize soil or compost, it is far easier to use materials that are already sterile.

Mixtures containing peat moss are often added to media used for rooting such acid-loving plants

STEM CUTTINGS

Herbaceous cuttings are taken from non-woody plants such as perennials, houseplants, and tropical plants while they are growing.

Softwood cuttings are pieces of new growth snipped from woody evergreen or deciduous plants in early summer before the growth has started to harden.

Semi-hardwood cuttings are taken from those same plants later in the summer, after the wood has hardened slightly.

Hardwood cuttings are taken when the wood is dormant.

LEAF CUTTINGS

A leaf or part of a leaf is taken from either a woody or herbaceous plant. The leaf does not become part of the new plant, but new shoots and roots grow from it.

ROOT CUTTINGS

Some plants reproduce themselves readily from small pieces of roots which are cut from the parent and planted in early spring.

as azaleas and blueberries. Both are difficult to root, however, and the moisture-retaining peat sometimes causes them to rot before they have developed roots. A medium such as (5) used on top of a layer of peat moss helps to solve the problem. Rooting takes place in the top medium, and the roots then grow into the peat underneath.

Foil will hold ivy erect in glass.

Stir water to add oxygen to it.

Pot when roots are well formed.

ROOTING ENVIRONMENT

Providing the right conditions for cuttings to root well has long been a challenge to gardeners. Those in some sections of the world have a moist, warm climate with hazy sunlight which makes it possible to root cuttings easily outdoors. The rest of us must duplicate these conditions as best we can.

A warm temperature and high humidity are essential. Given these, some plants, such as mock orange, will form roots even in the air. Light is necessary, too, but not in as great amount as for starting seedlings.

If you want to grow only a small number of herbaceous, softwood, or semi-hardwood cuttings, the proper humidity can be obtained by filling a plastic pot with moist sand or a vermiculite-perlite mixture. After inserting the cuttings, cover them with a glass jar or a transparent plastic bag and sprinkle them once or twice a day. Place them in a window where they will get light, but no direct sun, and keep them warm, but don't let them become overheated. A round-the-clock temperature of between 75 and 85 degrees F. is best for rooting most cuttings.

Propagation Box

Some gardeners build a propagation box to root small numbers of cuttings. These resemble aquariums. In fact, some people use aquariums for this purpose. They make it possible to create a very humid climate in a small area. They can be any size, and covered with either glass or plastic.

Hotbeds are good places to start cuttings. Since they are sunk in the ground, bottom heat, possibly supplied by a soil cable set at about 75 degrees F., is usually necessary, even in summer. If you choose this location, cover the hotbed with a piece of burlap or white sheet to shield the cuttings from direct sunlight, or they will dry out too rapidly.

Cuttings also root well in greenhouses, because they easily provide the required heat. Because of the bright sunlight and sometimes excessive heat there, however, provision must be made to add extra moisture, or the leaves on the cuttings will quickly dry and wilt. Humidifiers, fog units, or mist systems are used to supply this moisture, because ordinary watering is not enough. As with growing seedlings, make sure the temperature does not fall too low at night.

To have cuttings that are in first-class condition, you must work with good tools. A sharp knife or razor blade should be used for taking cuttings from houseplants, and well-made hand pruners for other cuttings. The sharper the tools are, the better, since you want to injure the tissue of the plant as little as possible, and dull knives or clippers crush stems badly.

Take cuttings only from healthy, vigorous specimens. Plants can be made to produce larger cuttings and more of them if you prune them heavily and fertilize them well in very early spring of that same year.

Label your cuttings carefully with the name of the plant variety and the date the cutting was made. You might want to add, also, the type of rooting chemical used, and any other pertinent data that might help you in future propagation.

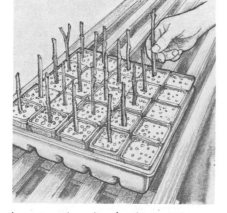

Insert cuttings in plastic containers.

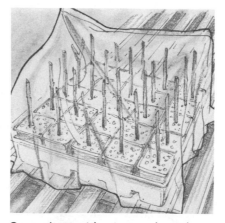

Cover them with a jar or plastic bag.

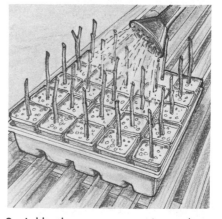

Sprinkle them once or twice a day.

ROOTING CHEMICALS

Although a few plants root easily, most benefit from the help of a compound that induces rooting. It is especially beneficial for softwood cuttings. Since they are taken in the summer, they must root quickly before they dry out or rot. Also, they need to develop a good root system early, so they will be able to safely withstand their first winter.

It is now a common practice for home gardeners, as well as commercial nursery people, to dip the bottom ends of cuttings in a rooting compound. Several chemicals have been developed that stimulate faster rooting of both hardwood and softwood cuttings. IBA (indolebutyric acid) is the most popular, and NAA (naphthalenacetic acid) is less used. Rooting compounds such as Rootone, Hormodin, Hormex, Jiffy Grow, Dip-and-Grow, and Hormo-root are usually liquid or powder forms of IBA, a mixture of both IBA and NAA, or some similar chemical. They are pro-duced in varying strengths and are usually classi-fied as being suitable for plants that are easy to root, those that are moderately difficult to root, and those very difficult to root. Most list on the label the plants for which they are recommended. It is important to use the proper compound, since one that is too strong may damage a sensitive cutting. Like many other growers, however, I sometimes use a stronger formula on woody cut-tings that I take in late summer, because at that time ordinary shrubs often root better with a stronger chemical.

Mix a tiny bit of Ben-Late or Captan into the rooting powder before dipping the cut end of the new cutting in it. A fungicide helps to prevent any disease the cuttings might pick up in the warm, humid atmosphere in which they will be living. After dipping, tap off any excess powder, since it is expensive and only a small amount is necessary and beneficial.

HERBACEOUS CUTTINGS

Many houseplants, perennials, vegetables, and tropical plants can be started from cuttings. It is best to take them at a time when the plant is actively growing, but many will root easily at any time, even without the use of a rooting com-pound.

We have successfully started such plants as bleeding heart, delphinium, dahlias, pinks, petu-nias, tuberous begonias, most herbs, and many houseplants, and have even increased our supply of hybrid tomato plants in this way. One year we were able to start very few of our favorite White Cascade petunia plants from seed. We took cut-tings from them, and by summer we had the three dozen needed to fill our window boxes.

Unless the plant is one that has naturally juicy sap, such as aloes and some cacti, water the plant a few hours ahead of the time you plan to take the cuttings, so it will be full of water. When you are working with a young plant, the cuttings may be taken anywhere there are sprouts long enough to be removed without wrecking the plant. If the plant is old, choose a sprout that grew recently, and if possible one that is pres-ently growing. Herbaceous cuttings may be of various lengths, but are usually from three to six inches long. A razor blade or sharp knife is a good cutting instrument, and is better than scissors, fingernails, or even hand clippers, which tend to crush the tender tissue. A smooth, clean cut will form a callus and heal better than a ragged one. Cut the bottom end on an angle to allow more surface for rooting. Pinch off all flowers, flower buds, forming seeds, and all except one or two small leaves from the cutting before inserting it in the medium.

Callus Forms

Often a fleshy growth will form on the cut end. This is the healing over of the cambium layer and is called a callus. The forming of a callus often, but not always, precedes rooting.

It is possible to root most herbaceous cuttings in a variety of media, including soil, sand, ver-miculite, perlite, moss, and sometimes even wat-er. My favorite is a mixture of half vermiculite

When taking a cutting, cut off the three- to six-inch section with a razor blade or sharp knife, then remove most of the leaves and any flowers or flowering buds. Don't let the cutting dry out before planting it.

and half perlite because it gives good results with less chance of promoting rot than most other combinations.

Treat the cutting with a rooting chemical, if one is recommended for that variety of plant (see Part Two). Most cuttings should be stuck into the medium as soon as they are made, so they won't dry out. A blunt pencil or a dibble is a handy instrument to make a hole for inserting the cutting. Don't try to insert it without making a hole first, or you may bend or damage the cut end. Insert it upright so the bottom end is about one inch below the surface of the medium. Keep the medium moist, but not wet, provide a temperature of from 70 to 80 degrees F., and keep the cuttings out of drafts.

Because it is essential to keep the cuttings from drying out or rotting before they root, the faster the rooting process can take place, the less chance there is for failure. Chrysanthemums, ivies, and many others root rapidly with little fuss. Others take longer and need more attention. Use a rooting chemical, and provide the proper humidity by covering the cuttings with a plastic bag or other means, as suggested earlier in this chapter.

Plants with a heavy, sticky sap content such as geraniums, pineapples, cacti, and some others, are less likely to rot before they root if they are spread out on newspapers in the shade and allowed to dry for an hour or two before they are inserted in the medium.

LEAF CUTTINGS

Certain woody and herbaceous plants start easily from leaf cuttings. Plants such as the African violet, sedums, hens and chickens, artillery plant, and peperomia can be increased by simply cutting off a leaf with its stem attached, and burying the stem in a medium. Either a mixture of vermiculite and sand, or vermiculite and perlite works well. Roots will form and a new plant will grow.

Some plants seem so anxious to reproduce in this manner that the leaves form little plantlets right on the stems. The well-known pick-a-back, or piggyback plant, does this on runners which can be removed and transplanted directly into a sand-vermiculite mix where they grow easily. Certain water lilies, air plants, kalanchoes, and spider plants act in a similar fashion by rooting and forming new plants wherever they touch the soil.

Leaf cuttings can be taken whenever there are fresh, full grown leaves on the plant. The cutting must be kept moist at about 70° F. until it begins to grow roots and a sprout, so it should be covered with a clear piece of plastic. Under these moist, warm conditions, it is sometimes difficult to prevent rot. As a precaution, keep everything as sanitary as possible by using a sterile medium and clean tools, and try not to bruise or otherwise damage the leaves when you handle them. An occasional sprinkling of a good fungicide such as Captan or Benomyl helps to prevent fungus infections.

The length of time it takes for roots to form varies with the plant, but they should root within a month or slightly more. After roots and new upward growing sprouts have formed on the leaf, it can be potted. Occasionally several plants start from the same leaf. These can be cut apart, and each individual transplanted into its own pot. Long, narrow leaves such as those of the sansevieria may be sliced into one-inch pieces and stuck vertically in the media. Keep the pieces top side up. The large leaves of some plants such as Rex begonias can be cut into ¾-inch squares and planted vertically the same way.

Surprisingly, some plants that root nicely from leaf cuttings will not make any top growth afterwards. The jade plant *(Crassula argentea)* and the India rubber fig *(Ficus elastica)* are two of these. Be sure to include a portion of the older stem with cuttings from these plants so that they will form a complete plant.

SOFTWOOD CUTTINGS

Softwood cuttings, the soft, growing sprouts taken from woody plants, are similar to herbaceous cuttings in many ways. Flowering shrubs head a long list of plants that may be started from softwood cuttings. I have had good success in rooting small fruits, grapes, elderberries, bush cherries, currants, and gooseberries, and moderate success with the hard-to-root blueberries and saskatoons. I have also rooted many trees from softwood cuttings, including weeping willows, hybrid poplars, yews, arborvitae, junipers, mugho pines, and flowering crabs.

Young, healthy, vigorous plants that have been pruned hard the previous season provide the best cuttings. Generally new roots form better if the entire cutting is made up of new growth, but on some plants such as hydrangea, potentilla, and similar ones, a bit of older wood may be included to make a larger cutting.

When To Take Cuttings

Late spring and early summer are the best times to take cuttings. Plants are making their fastest growth at that time, and the potential for root growth is the best, also.

Take softwood cuttings after a rain or a few hours after the plant has been well watered, so it will be in a turgid condition. No cutting should be allowed to completely dry out, but softwood cuttings are especially perishable. They are best taken in the cool morning, and kept in water for a half-hour or so, before they are stuck into the medium.

In making a softwood cutting, cut it from the plant at an angle, rather than straight across. As with herbaceous cuttings, a larger cut surface provides more area for it to callus and root.

How Long?

Experts argue about the ideal length for a softwood cutting. Europeans have traditionally favored short cuttings, but one that is six to ten inches has been found to root better than one three or four inches long, provided there is that much new growth on the plant. A great deal more reserve energy is stored in a larger branch, which aids initially in fast, heavier rooting, and also makes the cutting grow faster once it has rooted. Occasionally, as an experiment, I have rooted cuttings that were two feet long, and they grew rapidly into sturdy, rugged plants that were saleable the same year. The drawbacks are that far fewer cuttings of that size can be taken, and fewer can be started in a limited space.

Although much research has been done with softwood cuttings to find whether they root better if the cut is made at a node (leaf joint), below it, or just above, with the advent of mist systems, plastic-covered propagation chambers, and other improvements, the location of the cut is considered less important. In some experiments, weigela and common privet cuttings were found to root best when taken a half-inch above a node, but boxwood, cotoneaster, ginkgo, beautybush, potentilla, pyracantha, and buckthorn were among those that rooted better when cut right at the node. Aronia, barberry, dogwood, deutzia, forsythia, witch hazel, hydrangea, hypericum, kerria, leucothoe, honeysuckle, magnolia, apple, plum, currant, ninebark, spirea, and most viburnums rooted best when cut one-half inch below the node.

After dipping the cuttings in the rooting powder, stick the cut ends in the medium to a depth of from 1½ to 2 inches. Use a stick or dibble to make the holes so you won't rub off any powder while pushing them in.

I prefer a vermiculite-perlite mix for rooting most softwood cuttings. In the early days I used a medium of sand mixed with a little peat, but I find that the sterile artificial mix gives a lot less trouble with disease.

Use Plastic Pots

Originally, I started all cuttings in large flats. This system presented a problem, however, because some varieties of plants, such as potentilla, rooted very rapidly, while others, like blueberries, took most of the summer to root. Now I place all cuttings in small square plastic pots set in trays (available from nursery supply houses). This makes it possible to remove any one from the mist as soon as roots have formed. I learned early not to leave a cutting under the mist system after the roots formed, or it would soon collapse.

As with herbaceous cuttings, warmth and humidity are more important than bright light. They need only a moderate amount of light to root. One grower I know roots everything in a tight plastic tent on the north side of a large barn which gets almost no sun, but plenty of light. A soil cable with a thermostat under the beds provides heat at night and on cool days. Before he puts his cuttings into pots and flats, he dips the tops of each in Wilt-Pruf, an anti-transpirant that helps keep plants from drying out (sold in most garden supply stores). That treatment, plus his tight plastic house and regular waterings, supply the right conditions for rooting, and every year he gets good results.

Another amateur horticulturist acquaintance starts a wide variety of plants, from clematis to smokebush, by placing his cuttings in his basement under a battery of grow lights covered with a plastic tent. Heat, humidity, and light are easy to control in his indoor location, and, although his electric bills are high, his plant production is impressive.

THE MIST SYSTEM

I have found a mist system set up in a greenhouse most conducive to the successful rooting of cuttings. Mist makes it easy for anyone to propagate since the systems are now being designed for use by hobby gardeners. Garden magazines and catalogs offer small kits suitable for starting a few dozen plants easily, or you can buy the components and put together your own unit, making it as simple or as automatic as you want.

When mist systems were first developed, they consisted of a nozzle that sprayed a fine mist over the cuttings throughout the day. The water was turned on in the morning and off at night, by hand. Although it proved a huge improvement over all previous methods and was satisfactory for the plants that rooted fairly fast, those that took a long time to root often got too much moisture, especially on cloudy days, and many rotted. Today the amount of moisture can be regulated by devices that range from a simple timer which turns on the mist at regular intervals, to complicated sensors that measure the moisture and adjust the mist accordingly.

Tried Many Methods

Over the years I have tried various ways of misting cuttings. Although my first system was rather crude, it worked well and enabled me to start a large number of plants in a small, homemade plastic greenhouse. I bought nozzles from a nursery supply catalog and acquired a solenoid valve, to turn the water on and off, from a junked automatic washer, which I wired to a timer. I stayed nearby on cold or wet days to shut off the water so the plants would not be soaked with cold water for hours.

Before the season was over, I began to have

trouble because the fine holes of the mist nozzle plugged, even though I had installed a line strainer. I finally found it was caused by algae growth in the warm water inside the plastic pipe.

Later I used a mistblower sprayer for a short period, hoping it would be more reliable. The amount of water sprayed over the plants was not constant, however, and some plants got far more than they needed. Also, it was so noisy that we couldn't hear the telephone or much else when it was running.

I tried, too, a mist control with a so-called electronic leaf that determines the amount of water on the leaves and turns on more when it is needed. Although this machine apparently works well for many people, we had trouble with it. Because our water is very alkaline, as it evaporated off the fine screen which does the sensing, lime remained and weighed down the screen so much that the switch stayed off, and no spray came on.

Present System

The system we use now is very satisfactory. Our outside water supply is connected to a coil of 100 feet of black plastic pipe hung on a rack in the top of our greenhouse, where it gets very warm. The pipe is connected to a valve which can be shut off at any time, so we can work on the system. Near the valve are a line strainer and a solenoid valve. The flexible plastic pipe connects to one of rigid plastic that extends the length of the greenhouse and is hung about six and a half feet above floor level. Along this pipe I drilled holes at six-foot intervals and installed greenhouse spray nozzles. These spray the cuttings as satisfactorily as the mist nozzles, and do not become plugged nearly as easily.

One timing device turns on the electricity every morning at a certain time and turns it off in late afternoon. The electricity activates another timer which controls the solenoid valve. Every

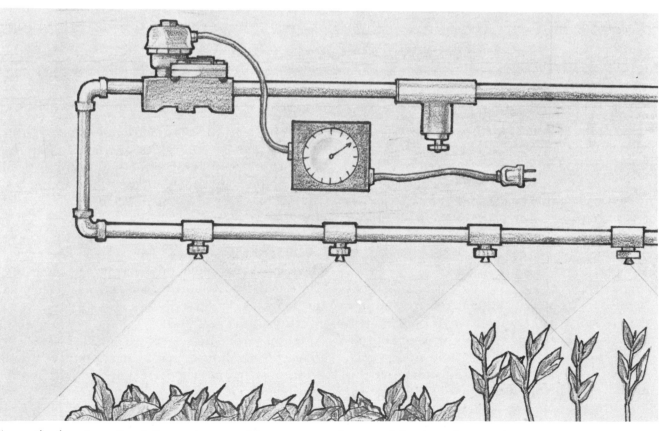

The author's automatic misting system provides a two-minute spray every ten minutes during the day. It is controlled by an electrical timing device. The system includes 100 feet of plastic hose, placed in a rack in the top of the greenhouse where the water is heated. The water is sprayed over the plants from a rigid plastic pipe which extends the length of the greenhouse.

ten minutes, the cuttings are sprayed with warm water for two minutes, which is about right for a sunny day. On cool, cloudy days, I shut the system off for a few hours.

Our mist system is set up in the greenhouse in early summer, after we move out the annuals that have been growing there during the spring. The cuttings root well in the warm early summer days, and as long as they never dry out, they stand a lot of heat. Since we live in a cool part of the country, however, it is necessary to close the greenhouse ventilators at night, so the heat will be retained until morning. It is important that the plants be warm at night as well as during the day, or a lot of valuable rooting time will be lost. Most plants root within three to six weeks.

Problem Plants

Although I have had great success in starting under mist many plants that I had found difficult to propagate in other ways, I discovered that others did not root well in a moist atmosphere. French lilacs, azaleas, amelanchier, certain evergreens, and others tend to either rot or drop their leaves or needles before they root.

Although these hard-to-root plants continue to present a challenge, I have found it helpful to use one or more of the following methods:

(1) After a week or two under regular misting, increase the time between sprays, so the leaves dry out slightly between waterings.

(2) Avoid misting entirely. Keep the media moist, but the leaves dry most of the time. A propagation box often works better than a mist system for these kinds of plants.

(3) Use a medium that holds little moisture, such as coarse perlite alone. This is particularly useful with slow-rooting evergreens. One grower we know has even successfully rooted balsam fir cuttings by using peastone as the media.

(4) Instead of softwood cuttings, try rooting the plant by taking semi-hardwood cuttings.

SEMI-HARDWOOD CUTTINGS

Some plants root best when their cuttings are taken after the new growth has partly matured, in late summer. Woody plants, such as broadleaf evergreens, and deciduous shrubs such as flowering currant, deutzia, forsythia, red leaf plums, weigela, and certain dogwoods are in this class. The cuttings should be taken in the same way as softwood cuttings, and placed in sand or another medium in the same way. Rooting chemicals of the stronger type are usually recommended for cuttings taken in the late summer, fall, and winter months.

Semi-hardwood cuttings may be rooted in a greenhouse, propagation box, or, if the climate is mild, in an outside cold frame. Large-scale propagation, however, is always done under mist. Because the days are cooler and shorter at this time of year, bottom heat must be supplied for the entire rooting period. If mist is not used, water should be applied frequently enough to keep the medium moist and the tops from becoming dry. Bottom heat has a drying effect, so frequent checking is necessary.

Lighting

It is difficult to say for sure how much light is enough, but, in addition to daylight, after the days shorten, several extra hours of artificial light, either incandescent or fluorescent, are almost necessary for the good rooting required for healthy plants.

By late winter, when good root systems have developed and the tops have started to grow, the days should be long enough so no additional light is needed. By late spring, the cuttings should be moved out of the greenhouse. (See After-Care of Cuttings/next section.)

If you garden in a warm part of the country or have a greenhouse that is heated year-round, you may want to experiment with semi-hardwood cuttings.

Because of the expense and extra problems connected with coping with cold temperatures and short days, most amateur home gardeners in cool regions leave this method of propagation to commercial growers.

AFTER-CARE OF CUTTINGS

The first year with a mist system, I was delighted when our cuttings rooted so successfully. The success was short-lived, however. When I moved them outdoors, most of them promptly wilted and died. The change was too much of a shock for the pampered plants. It became obvious that a conditioning period was necessary between the damp, hot greenhouse and the natural world of sudden temperature changes and variable moisture.

Now we adjust them to their new environment in several stages. When the cuttings are first taken out of the greenhouse, we put them in a heavily shaded, fairly warm, plastic-covered house. They are given one feeding of liquid fertilizer, to help lessen the shock of the environmental change, and I water them each day.

First, plants are placed in this plastic-covered house.

The roots continue to develop in the shade house, and after about a week they have become adjusted to not being misted. I then move them outdoors, but give them protection from the sun by placing them under a tent of woven plastic or burlap. The covering extends about six feet above the plants, and over the south and east ends. Rain can penetrate the shade cover, and although the cuttings still get adequate light from the open north and west ends, they are protected from the wind and hot sun.

We continue to water them, and once during their week there, feed them with liquid fertilizer. Then they are moved outdoors where the watering and once-a-week feeding is continued. They not only survive, but thrive, as long as they are not allowed to dry out. On hot, windy, dry days, they must be tended carefully.

After a week, they're moved outdoors, under plastic tent.

As soon as their roots fill the pots, we plant them outdoors in beds or transplant them into larger pots where they will grow until they are sold.

Even after being transplanted, they will need watering and feeding. Discontinue all feeding after August 1, so the plants don't make too much late summer growth. The plants in outdoor beds should need no winter protection, but potted ones may need to be covered or moved into a shelter covered with white, not clear, plastic, unless you live where temperatures are mild, or snow buries the plants for the winter.

Final step is to pot them or to plant them outdoors.

HARDWOOD CUTTINGS

As the name implies, hardwood cuttings are taken when the wood is dormant. Most woody plants were started from hardwood cuttings until the introduction of mist propagation, and quite a few are still grown by this method. Because these cuttings require very little equipment, it is a convenient way for home gardeners to propagate. We have successfully started grapes, currants, willows, spireas, cotoneasters, shrub roses, hydrangea, honeysuckle, mock orange, privet, ninebark, numerous vines, and certain evergreens from hardwood cuttings.

The cuttings can be taken anytime from late fall until late winter. Usually it is best not to take them too early in the fall because some plants, such as yews, need a chilling period of several weeks of cool weather in order to root well. We usually take ours in early winter, just before the snow comes.

It is interesting to note that the place the cutting is taken on a yew may determine the new plant's future growth habit. Cuttings taken from the top will usually grow into upright, cone-shaped plants, while those taken on the sides may become spreading, dwarf or globe-shaped. Because of this peculiarity, many "new" varieties of yews have been introduced that probably aren't new at all, and cannot be reproduced accurately.

It makes no difference in the future growth habit of most plants, however, where the cutting is taken. Select them only from healthy wood that grew the previous summer, and reject any branches that look weak or diseased. It is possible

Take cuttings when wood is dormant. Make all cuttings the same length.

Tie into small bundles, then bury them in a can of damp vermiculite.

In the spring, dip the callused bottom ends into rooting powder.

Plant so that only top buds remain above ground. Cover with plastic.

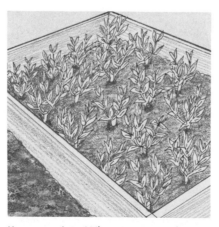

Keep moist. When tops and roots have formed, remove the covering.

Overwinter in the bed, or, in warm climates, transplant.

to get several cuttings from the same branch of some fast-growing, deciduous shrubs.

Cut the top end on a slant, slightly above a bud. Cut the bottom end on a slant also, to expose more cut area to form roots. I try to make all the cuttings of each variety the same length, between five and twelve inches. That makes it easy to tie them in small bunches, and I always make sure that all the top ends are placed in the bundle in the same direction so it will be easy to stick them in the soil right-side up. Label each bundle carefully with plastic labels marked with waterproof ink or an indelible pencil. With the identifying leaves gone, the bundle of twigs looks quite anonymous, like unpainted pencils.

Bury the bundles in a container that holds enough of the medium to cover them adequately. We use slightly moistened vermiculite in large plastic garbage cans, and store them in our root cellar, which stays cool but does not freeze. Some people use sand, but I like to use vermiculite because it is lighter and easier to manage than sand, and is less likely to harbor disease. Whatever you use should be barely moistened and never soaked. Toward spring, when the cellar starts to become warmer, the cuttings begin to form a thick callus on their bottom ends. Wait until they have developed the callus before setting them out, however. If they are set out too early, a few leaves may form, but roots will not develop in time to support them.

In my early gardening years, I planted the callused cuttings directly in the ground, without giving them any special care. Although many rooted and grew, the losses were high. Now, before planting, I dip the callused ends in a rooting powder. I plant them either in a hotbed or an outdoor bed of well-prepared, light, rich soil in a protected spot that gets sunlight only until noon. The cuttings are placed four to six inches apart and each is buried deep enough so that only the top one or two buds remain above the ground. I firm the soil down around the cutting so it won't dry out easily, and cover the entire bed with a sheet of slightly cloudy polyethylene, supported a few inches above the cuttings by boards surrounding the plot.

Whenever the weather is at all dry, I water the bed and continue to water it even after the roots have started to form. Until they have developed a good root system they cannot survive if the soil becomes dry. As soon as the roots have started, I give them some liquid fertilizer such as Peters 20-20-20 or Rapid-Gro.

When the new plants have developed good top and root growth, I remove the plastic cover. Thanks to the rich soil and regular feedings and waterings, they grow into sizeable plants before the end of the summer. I leave them in the bed over the winter, and either pot them or transplant them into the field early the following spring.

In mild climates they may be transplanted in late fall.

ROOT CUTTINGS

Root cuttings are much like stem cuttings, except that the root, or part of it, is used rather than the branch. Root cuttings are not one of the most popular methods of propagation, but they are useful for starting certain plants that are difficult to propagate in other ways. Nurserymen sometimes start dwarf apple and plum trees from root cuttings. They are grown for two years and are then used as rootstocks for grafting.

Although I have had good luck starting large numbers of phlox, flowering quince, oriental poppies, and lilacs by root cuttings, I have found better ways to start most varieties. Commercial propagators start certain perennials and berry plants in this way. One nurseryman friend of ours grows large numbers of raspberry and blackberry plants each year. In the spring he digs entire plants, cuts their roots into pieces about an inch long, and plants them about two inches apart in a bed he has rototilled and enriched with lots of manure. Although he loses the parent plants, within a few weeks thousands of shoots come up, and by fall, with his careful nurturing, they have become plants that are big enough to sell.

When taking root cuttings, dig up plant, and cut off roots, making slanting cut on bottom of roots.

If plant is to be replanted after cuttings are taken, the top should be pruned to reduce load on roots.

Depending on the type of plant, roots can be started in flats or a variety of other containers.

Work in Early Spring

Early spring is the best time to take root cuttings. Although a few small pieces can be chopped from the roots of a plant in the ground without disturbing it too much, the usual method is to dig the entire plant, cut the roots into pieces, and either replant the remainder of the old plant or throw it away. If it is replanted to grow more roots, the top must be cut back severely, so there will not be more top than roots.

The part of the root that is closest to the main root of the plant is considered the top of the root cutting. This is important to remember because some plants must be planted vertically and right-side up. Propagators often make their cuttings with a straight cut across the top and a slanted one on the bottom so they can make a speedy identification while planting them out.

Root cuttings do best if they are planted in light, loamy soil that is rich in organic matter and nutrients. They are usually started in the ground, but if you want only a few, they may be planted in pots or deep flats. Give them plenty of sun and water them whenever necessary so the shallow roots never suffer from drying out.

There are three primary methods of starting root cuttings:

(1) Perennials with fine roots can be started easily in large numbers by cutting the roots into pieces one to two inches long, and scattering them over the surface of well-prepared, rich, sandy soil. Cover them with a half-inch of sifted soil or sand and keep it moist. Achillea, crownvetch, sea holly, flowering spurge, gaillardia, soapwort, phlox, perennial salvia, stokes aster, and verbascum are only a few of the plants that root and form tops easily by this method.

(2) Plants with fleshy roots, such as acanthus, old-fashioned bleeding heart, peonies, baby's breath, beebalm, rhubarb, and Oriental poppy prefer a different treatment. Make the cutting slightly longer, from 1½ to 2½ inches. Make peonies three inches long. Plant them vertically, top side up, about three inches apart, with the top end of the cutting protruding about a quarter of an inch above the soil.

(3) The root cuttings of many woody trees and shrubs reproduce best when treated as in (2), by planting the root piece vertically, but with the top 1 or 1½ inches *below* the surface. The cuttings should be large—from 4 to 6 inches long.

Included in this group are apples, plums, peaches, and cherries (all grown for their root-stocks), aralia, artichoke, blackberry, cork tree, crape myrtle, dewberry, elderberry, flowering quince, golden rain tree, horseradish, hypericum, lilac, locust, mountain ash, roses, saskatoon, sassafras, shadbush, wisteria, yucca, and numerous others.

Root cuttings seldom start to grow quickly, so you must be patient as you await the first sprouts. Careful handling in making and placing the cuttings, as well as good after-care, will help

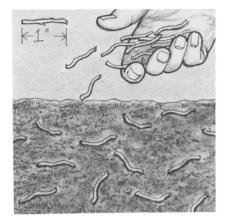

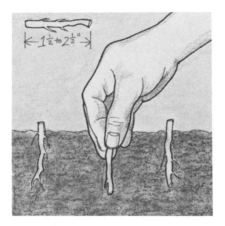

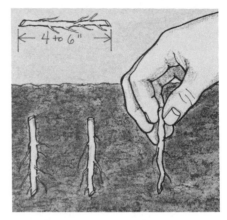

Cut roots of fine-rooted plants into one-inch pieces, scatter them on soil, and cover with light layer.

Plant fleshy roots, 1½-2½ inches tall, top side up, and with a quarter of an inch protruding from the soil.

Bury the top of the root 1-1½ inches when planting the roots of woody trees and shrubs.

ensure success with root cuttings. There will be few years, however, when you can count on 100 percent of them to root and grow.

There are exceptions to nearly every rule, and the statement that plants started asexually will be exactly like their parent does not always apply. Root cuttings from thornless dewberries nearly always produce plants that have thorns. Likewise, some varieties of geraniums, bouvardias, and a few of the plants with variegated or colored foliage revert to more common varieties when started from roots.

AIR-CALLUSED CUTTINGS

An air-callused cutting is a combination of an air-layered and a softwood cutting. The general idea is to induce a callus to form while the potential cutting is still attached to the parent tree or bush. Then, after the cutting is finally made, it roots quickly and more heavily.

Air callusing is used successfully on plants that are difficult to root by regular cuttings, such as Japanese tree lilac, blueberry, apple, plum, flowering crab, cherry, pear, daphne, azalea, rhododendron, and a variety of broad and narrow-leaved evergreens. It is used also to induce grafted plants, such as fruit trees or French lilacs, to grow on their own roots.

There are many modifications of the air-callusing method. Because it is a fairly new development in propagation, improvements are continually being made. No one method has yet emerged as the best, so you may want to experiment with procedures of your own.

The basic steps:

(1) In early spring, find a suitable branch on a dormant tree or shrub of the variety you want to propagate. The future cutting should consist of the previous season's growth, and be no larger than a pencil in diameter. About eight inches from the top end, cut a strip of bark from ½- to ¾-inch wide from the cutting-to-be, completely encircling the branch. Lift it carefully and completely from the wood, being careful not to cut into the wood itself.

(2) With a small watercolor paintbrush, cover the bare wood on the branch with a thin coating of paste you have previously made by mixing one part extra-strength rooting powder with an equal amount of Captan (to control disease), a dash of confectioners' sugar to help supply plant energy, and enough liquid growth retardant, such as B-Nine, to make a paste. The growth retardant is used by bedding plant growers to get seedlings such as petunias to grow short and bushy instead of tall and straggly, and is available

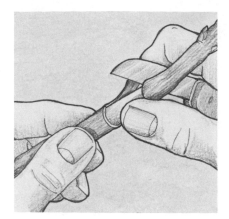

Cut a half-inch circle of bark from an air-callused cutting.

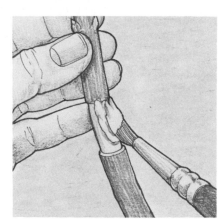

Cover bare wood with mixture of Rootone, Captan, and B-Nine.

Cover this with sphagnum moss and plastic, tying them in place.

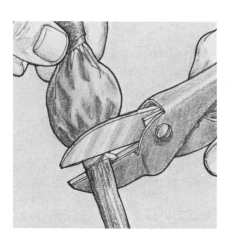

When callus has formed on wound, cut off cutting below the wrapping.

Unwrap it, dip end in rooting powder, then plant it in a pot.

If possible, keep this new plant under a mist until new roots form.

from nursery and greenhouse supply houses. When spread on the wound, it apparently counteracts certain rooting inhibitors that are present in the tree.

(3) After the wound has been covered with paste, surround it with a wad of slightly moist sphagnum moss or a split Kys-Kube, a fiber block that is also available from nursery and garden centers. Then, with a plastic food wrap or Pliofilm, wrap it carefully, making it as moisture-tight as possible. Fasten the wrap with plastic electrical tape or a plant tie. Cover the plastic wrap with a piece of aluminum foil to keep out the hot sun. Crimp the foil tightly together to keep it in place.

(4) In one or two months, a thick, fleshy callus should have formed on the wound. The callused cutting may then be clipped off, just below the wrapping.

Unwrap it, dip the callused end in rooting powder, stick it carefully in a pot containing a mixture of moistened perlite and vermiculite, making sure the callus is well buried. From now on, treat it as a softwood cutting. If possible, keep it under a mist until roots form. Presto! A plant with no grafts to worry about.

GRAFTING

When the back-to-natural-living movement was in full swing in the late sixties, a new homesteader in our town called to discuss fruit trees. He asked how long it would take to grow a producing orchard if he planted seeds.

I explained that trees grown from seeds rarely bear good fruit, and that he would have to graft his trees if he wanted to grow any fruit that was worthwhile. He said he felt that grafting was an unnatural practice and he wanted his orchard to be completely natural, even if it meant growing only small, hard apples. I explained that grafting was merely the transplanting of a good variety of fruit on a poor variety, and I had never felt it was unnatural or harmful. I pointed out that grafts appear in nature occasionally when two adjacent trees grow together. He remained unconvinced, and set forth to grow his fruit from seed.

We never knew how the project would have turned out, because his goats, also doing what comes naturally, ate the trees. He abandoned the orchard, and finally gave up farming and returned to the city.

The Mystery of Grafting

Grafting was treated as a mystery when I was young, and no one knew much about it. Nearly everyone skilled at grafting had passed on, and they seemed to have ensured their reputations by not telling anyone else how to do it. Because I was curious I asked a lot of questions and heard a lot of unusual stories.

Some remembered an elderly man who went from place to place in the spring, grafting wild apple trees for a few pennies. One old farmer was sure beeswax was involved, but he didn't know how. Another said his father had grafted several apple trees successfully by plastering the graft union with cow manure and clay. Still another recalled that in his youth he'd seen one tree upon which a dozen varieties of apples, pears, plums, cherries, and even a few tomatoes ripened simultaneously on adjoining branches. "A good grafter," one man told me, "can put a graft on a full-grown wild apple tree, and the following year it will bear bushels of the best possible fruit." Then he added sadly, "Of course only a few people are born with such a gift." With such fascinating tales floating around, it is no wonder that I decided to try it myself.

Since none of the oldtimers could help me, I learned how to graft from a book, and to the surprise of my neighbors, who still preferred to think of grafting skill as a gift from the gods, my first results were good. Grafting, and its companion bud-grafting, or budding as it is commonly called, is only another type of asexual plant propagation.

One dictionary defines grafting as "the joining of two separate plant parts, such as a root and stem or two stems, so that by regeneration they form a union and grow as one plant." It adds that the word originates from Greek grapheion, meaning "to draw or write." Someone no doubt originally likened the shape of a shoot for grafting to that of a stylus or pencil! There are many good reasons for the grafting we do on fruit trees, named varieties of shade and flowering trees, roses, special kinds of evergreens, certain grapes, a few varieties of houseplants and certain perennials such as the Bristol Fairy baby's breath.

(1) The new plant will produce flowers or fruit like its parent. Horticulturists discovered long ago that when they found a superior wild plant, they often couldn't reproduce it by planting its seeds, because the seedlings failed to inherit its desirable characteristics, and reverted instead to a type that resembled its wild ancestors. By grafting, they were able to reproduce the superior plant itself.

(2) The size and shape of a tree or shrub can be controlled by its understock. For instance, dwarf apples, pears, cherries, and others can be produced by grafting scions from ordinary, full-sized trees onto special roots that cause them to grow as small trees. By using different rootstocks, a Winesap apple can be grown as a dwarf apple, about six feet tall, a semi-dwarf of nine to twelve feet, or a full-size tree that might grow to twenty feet or more. Special shapes can also be brought about by grafting. Such plants as tree roses and weeping birches are produced by grafting the desired top on an upright rootstock.

(3) Multiple grafts on the same plant are possible. Several different kinds of apples, or different colors of flowering crabs or lilacs, may be grown on the same tree or shrub.

(4) The vigor of certain plants can be increased. If a plant has been over-hybridized and makes weak growth when it is reproduced on its own roots, it can be grafted to more energetic roots. Tea roses, for example, are usually grafted on a multiflora rose or similar stock to help them grow more vigorously.

(5) Trees can be made to bear fruit sooner. Grafted fruit trees and nuts bear much earlier in life than those grown from seed. This characteristic is useful, not only for orchardists but also for experimenters, who sometimes graft a limb from a young fruit tree that they have grown from seed onto a larger tree to speed up its bearing. It is possible, in this way, to test a great many different seedlings on one tree.

(6) Certain plants can be made to adapt to unfavorable soil conditions. Peaches are often grafted on plum roots, so the trees will grow in heavier, cooler soils than the peach roots require.

(7) Grafting makes it possible to produce quickly a huge number of plants from one par-ent. This is useful, for example, when an unusually desirable variety is introduced and is in big demand while stock plants are scarce.

(8) It is possible, by grafting, to change the variety of a tree. If, for example, your Jonathan apple starts to bear, and you decide you don't like the flavor of Jonathans, you may regraft the young tree with some other apple.

(9) Grafting is useful if your space is limited and you need two fruit trees for cross-pollination. If you have only room for one, you can graft a limb of another variety on your present tree.

Drawbacks of Grafting

Although grafting is the best method of propagation for many plants, it has its drawbacks. As a young gardener, I was occasionally shown a special plant or tree, and the owner proudly said, "This is a grafted tree!" as if it merited certain respect. The mystique of grafting was so great that people often bought a grafted lilac or grape, when one growing on its own roots would have been better.

That trend has changed, and the art of grafting is not as widely practiced as it once was, for the following reasons:

(1) Plants may sometimes reject a graft, making a good union difficult or impossible. The rejection may take place the first year, or several years later.

(2) The graft union may be extremely fragile and subject to winter cold injury or to wind breakage.

(3) Many rootstocks used in grafting sucker badly. The vigorous shoots they send up from below the graft must be removed frequently or they will crowd out the good tree and eventually replace it.

(4) Grafted trees are often lost when mice or rabbits chew the bark and girdle the trunk. Any sprouts coming from below the graft will be from the wild understock. When a tree is not grafted, however, a healthy sprout growing from the roots can become a good tree with pruning and shaping.

Time To Graft

The time for outdoor grafting is determined by the fact that sap must flow from the roots into the newly inserted dormant scion as soon as the two are joined. Therefore, it is usually done in the spring.

Grafting indoors can take place at different times. So-called "bench" grafting is nearly always done in the winter, with dormant scions inserted in dormant rootstocks that were either dug in the fall and stored in a cool place, or purchased at that time. After grafting is finished, the completed grafts are packed in a moist medium and stored until they can be planted outside in the spring. Greenhouse grafting is usually done on herbaceous plants such as cactus and other houseplants, and on evergreens. It can be done anytime, but is usually more successful if the process takes place in late winter or spring, at a time when the plant is just starting to grow.

EQUIPMENT

Sharp, good quality tools are essential for a precision grafting operation, since neither torn bark nor wood heals properly. You will need a pair of hand clippers for collecting the scions, and a knife for making the grafting cuts. A folding blade knife is the most convenient and easiest to keep sharpened unless you have a protective holder for a fixed blade knife. If you plan to graft many large trees, a special, heavy, wedge-shaped knife called a grafting tool is available. A small hammer is used to pound it into trunks or limbs too large for splitting with a knife.

Each grafter has a favorite material for sealing the wound. One of the oldest and most often used is melted grafting wax, which is spread carefully over the wound with a small but fairly wide paintbrush of the type used by artists. It is always a problem to get the wax melted, especially if the grafting is done outside on a cool day. I have found that a kerosene- or alcohol-burning wax melter is almost a necessity if you are grafting many trees. The firms listed in the appendix that sell grafting supplies offer these wax melters as well as various waxes, tapes, and knives.

To save the mess of melting the wax, many amateur grafters use a cold wax. This type is supposed to soften in your fingers, but I have found it messy and hard to use on cold days, because it is not pliable. Many propagators prefer to use other materials to seal the graft union. Tree dressings such as Tree-Kote, grafting tape, rubber electrical tape, freezer tape, narrow strips cut from a thin plastic such as Saran Wrap, Parafilm, and others, work well. Whatever you use must be flexible enough to expand with the plant as it grows and not crack in cold weather or as it vibrates in the wind. It should also eventually deteriorate and fall off. When using a tape or strip, be sure not to wrap it with too many thicknesses, or it will not stretch and may girdle the plant. If a more permanent material, such as plastic electrical tape, is used, you must cut it away by early summer or it will constrict the growth of the tree.

GRAFTING FRUIT, NUT, SHADE, & ORNAMENTAL SHRUBS

Suppose you covet a nifty Delicious apple tree in your neighbor's backyard, and you find, in your own flower border, a small, two-foot tall apple tree that accidentally grew from a seed in the core someone threw out. On a warm day in late winter, catch your neighbor in a good mood, make him an offer he can't refuse, and under his watchful eye, snip off a small branch called a scion (pronounced sigh-on). By grafting, you can acquire a duplicate of his Delicious tree for your own yard.

Collecting Scion Wood

Normally a branch of pencil thickness or less is chosen for the scion (1/4" to 1/2"). The diameter of the scion wood you collect depends upon the

When cutting scions, use sharp clippers that don't crush the wood. Gather several weeks before grafting.

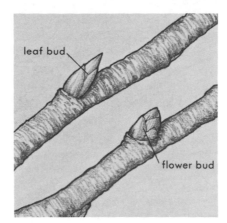

Collect branches with leaf buds . Leaf buds are pointed and narrow; flower buds are plump and round.

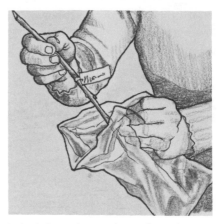

Label the scions, place them in an airtight plastic bag, and store them in refrigerator, not a freezer.

kind of graft you plan to make. For bark or cleft grafting, it is not essential that the scion be the same size as the rootstock, because only one side of the bark needs to be lined up with the stock for the sap to flow from one to the other. The scion should never be any larger than the stock, however. For whip grafting, the scion must be exactly the same size as the stock.

Gather the scions a few weeks before you intend to graft, if possible. They can be collected almost anytime during the winter when the wood is dormant, but not frozen. If they are stored more than a few weeks, it may take longer for growth to start. Snip them off with sharp clippers that don't crush the wood.

Collect only branches that grew the preceding year. On a young, vigorous growing tree, it is easy to find new wood, and new growth over a foot in length is usually best for a scion. It will be cut in much shorter pieces before you use it. Older trees grow more slowly, and you may have to hunt for some time to find good fresh twigs with fat buds. In my search for old apple varieties, I sometimes had to climb to the top of a 100-year-old tree to find good scions. Older trees can be lured into producing more grafting wood by pruning them moderately hard during late winter of the year before you want the scions.

Leaf Buds

Collect branches with leaf buds only, not flower buds. You can easily tell the difference by the size of the bud. The flower buds are plump

and round, and the leaf buds are usually narrow and pointed. Ordinarily the new growth at the end of the branch contains only leaf buds. Discard the one or two buds at both ends of the branch, and save only the wood with the nice fat ones in the middle. The terminal buds are too succulent and low in stored carbohydrates, and those near the base are more latent and likely to start to grow more slowly.

Label the newly cut scions, wrap them in a plastic bag, tie it so it will be airtight, and place them in the refrigerator (not the freezer) until you're ready to use them.

Larger amounts can be stored in a cold root cellar.

By gathering the scions ahead of time and keeping them cool, they will still be dormant when you put them on the branch. If you were to collect them at the time of grafting, the sap would have started to flow and the buds begun to swell, and the graft might not be able to support it.

For greenhouse grafting, the scions of hardwood and evergreen plants can be gathered at the time they are needed, as long as they are dormant. With soft pithy plants, such as cactus, the scions should only be cut immediately before they are to be used.

If you cannot find the varieties you want locally, you can buy scions by mail order from horticultural societies, fruit-testing organizations, and sometimes from private collectors. See the list in the Appendix.

CHOOSING A ROOTSTOCK

Although it would be fun to grow peaches, figs, cucumbers, grapes, and coconuts all on the same tree, such a practice is biologically impossible. To make a successful graft, the understock and the scion must be closely related. Apple varieties are grafted on apple rootstocks, pears on pears, and so on. The stone fruits—plums, peaches, apricots, nectarines, and cherries—may be grafted on each other, but not on an apple or an oak. As a general rule, the closer the scion and stock are related botanically, the better the chances for a successful graft.

Some surprising unions are possible. Lilacs are sometimes grafted on ash or privet, pears on quince, and a tomato plant can even be grafted on a potato plant. Gardeners are often delighted to find it is possible to graft pears on apples, and apples on hawthorn successfully, but usually the grafts live only a few years.

Sometimes Incompatible

It is difficult to understand why two varieties that should be perfectly compatible do not succeed. When some apple varieties are grafted upon certain apple rootstock, for example, the graft dies within a few years. Nurserymen and orchardists have long been aware of this problem, but it has been difficult to understand. With cherries and some plums it sometimes appears to be related to a virus that is present in either the scion or the rootstock. Some of these viruses develop faster in cool humid climates. Many grafted shade trees, such as the Crimson King maple and some varieties of prunes and sour cherries, are short-lived in northern New England, for example. Although they appear to be perfectly hardy and live for years elsewhere, in the North the graft union often separates before the tree is ten years old.

Influences Of Rootstocks

As I mentioned before, different rootstocks can be used to grow trees that will ultimately grow to different heights. The seedling growing in your flower bed would probably produce a full-sized tree, since this is the habit of most trees grown from seed. Seedling trees are usu-

ally the kind grafted by home propagators, as they are the easiest to obtain. Some gardeners plant a few apple seeds each fall. The seeds should start to grow the following spring, and if they grow well all summer, the trees should be ready for grafting the next spring.

The rootstock influences the tree in other ways, in addition to size. The flavor of the fruit and time of ripening are also affected. For this reason we don't dig up wild trees from roadsides or pastures to use as stock, because many of them grow from sour green apples. We now use only vigorous growing varieties, including some crab apples such as Dolgo. Many of our rootstocks come from seeds collected from the tasty apples that we use for making cider each fall.

Some nurseries sell apple, pear, plum, cherry, and other fruit tree seedlings for grafting. Some seed companies also sell fruit tree seeds so you can grow your own, if you wish (see Appendix). If you want to grow dwarf trees in large numbers, you may order Malling rootstocks of these, as well as other rootstocks, from Grootendorst Nursery, Lakeside, MI 49116. If you need only one dwarf tree, however, you may be able to dig a sucker offshoot growing from the roots of a dwarf tree, and transplant it to a good location. You will need to nurture it for a year or two until it looks sturdy before grafting it, however.

Double Grafting

Occasionally trees are "double grafted" or "double worked." The rootstock is first grafted with one variety, allowed to grow for a season, then grafted again with the variety wanted. Double grafts, or two-story grafts, are used when the interstock provides some characteristic that the scion or rootstock does not have. For example, dwarfing interstem might be placed between a seedling root and the final graft, or a disease or cold-resistant variety might be added. Interstock might also be used when there is a possibility that the rootstock and the desired graft may not be compatible. It can be used, as well, to modify the effects of the sour flavor contributed from a wild seedling root to the fruit of the grafted top.

THE GRAFTING OPERATION

On a warm spring day when the buds on the tree you are using for understock have begun to swell, you'll know that sap is running in the fruit trees and the time is right for grafting. Fruit tree sap starts later than the willows and maples, you'll notice.

First Steps

The day before you intend to graft, remove from the refrigerator the scions you have bought or gathered earlier, and snip off the bottom ends to expose a fresh area of the branch. Place them with their bottom ends in a pail of water overnight at room temperature, so they will absorb moisture and be less likely to become dry after grafting.

Grafting is a precision operation, and it should be done precisely, following careful sanitary practices. The scion must be attached to the tree with the cambium layers of both sections lined up exactly, so that sap will be able to flow easily from one part to the other. The cambium, which is the growing part of the plant, is the green layer that lies just between the inner and outer bark. Most grafting failures are caused because the alignment was not precise originally, or because the two parts were accidently moved out of alignment later.

Length Of Scion

The length of the piece of scion wood you cut depends on the type of graft you are using and the size of the limb into which it is being inserted.

For most grafts, a scion containing two buds is about right, but for small trees, it is better to use a piece with only one bud. Additional buds require that the graft transmit extra sap through the small cambium surface, and if it is unable to send a large enough volume, the scion is not likely to survive.

Cut On Slant

Cut the top of the scion on a slant about a quarter-inch above a bud, so there will be no dead stub left above the bud to dry out and rot. The top bud of the scion will grow into your new tree or branch.

For precise directions about how to join the scion and the understock, see the upcoming section.

Grafts are classified according to where they are placed on the tree (root, stem, or top), and by the method used in joining the two parts together (cleft, bark, veneer-inlay, whip, splice, saddle, and others).

Prevent Moisture Loss

The grafting wound must never be allowed to dry out and should be completely sealed with one of the materials mentioned earlier. The top end of the scion should also be covered with wax or tape to prevent moisture loss there. Sealing the wound not only keeps in the moisture, but it also helps prevent bacterial infection which can prevent healing.

When ready to graft, snip off the bottom ends, place in a pail of water.

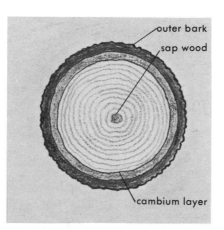

Cambium layers must come in contact for an effective graft.

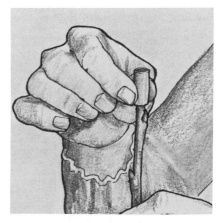

Plastic tubing over stub will supply moisture to the scion.

In addition to protecting the graft union, some grafters try to minimize evaporation from the scion, so the small amount of sap flowing to it through the joined cambium layers can support it. To do this, they cover the entire graft with a bag of white plastic (not transparent or black), or dip the part of the scion above the cut bottom end into melted paraffin or grafting wax before aligning the layers.

A newer method is to leave a slightly longer stub above the top bud when you cut the scion.

Then, after the graft is made, slip a short piece of plastic tubing of the same size over the stub above the bud and fill the tubing with water. The water supplies the scion with moisture from the top end, where the bud is, and further ensures the graft's success.

Unless the graft was placed below soil level, you will be able to see the spot where the grafting took place for several years.

Grafts may be placed at various positions on a tree or shrub, and are classified as follows.

ROOT GRAFT

A dormant scion can be grafted on a dormant root or piece of root. Usually this type of grafting is done during the winter.

Dig the roots in the fall and tie them in bunches. Store them in moist peat moss or vermiculite in a root cellar or some other place that is 40 to 45 degrees F. Apple, cherry, pear, and plum seedlings for root grafting can also be purchased in mid-winter from wholesale nurseries that specialize in them. (See Appendix.) The tops of these trees are cut off and the roots used as stocks.

Either the whole root clump or a piece of root three or four inches in length and about a quarter-inch in diameter may be used. When cutting root pieces, lay them out on the bench with their top ends facing the same direction, because it is important that the graft go on the right end. The bottom of each scion must be grafted to the top of each root. The top end of a root is the part closest to the main root of the tree, regardless of its position in the soil before digging. There seems to be no disadvantage in using pieces of roots as stock rather than the entire root, provided, of course, that the roots are cut from a vigorous tree.

Gathering Scions

Scions may be gathered any time during the fall and winter when the wood is not frozen. Select branches that are the same size as the rootstocks they will be grafted upon. They, too, may be stored in moist peat or vermiculite in a cool root cellar, or tied in a plastic bag in a refrigerator.

Grafting is usually by the whip method, although small amounts may be cleft-grafted if they are handled with great care after grafting.

Securing The Union

After the graft is made, the union of root and scion is secured with waxed grafting string, or thread, as it is sometimes called, instead of wax, to make a stronger unit. Sometimes the entire scion and graft, but not the root, is dipped in melted grafting wax or paraffin to prevent further drying out.

The new grafts should then be tied in bundles to prevent any unnecessary movement, and once again stored in peat moss or vermiculite in a cool place to await spring planting. The grafts are fragile at this point, and must be handled very carefully and as little as possible.

In the spring, after the ground warms, the grafts should be taken out of storage and planted either in the field in well-prepared soil, or in pots that are eight inches deep or larger, with the graft about an inch below ground level.

After growing a year or two, they may be transplanted to their permanent location.

Preferred By Many

Root grafts are preferred by many orchardists who grow standard, full-sized trees, because it is possible to plant the young tree deep enough so new roots will grow from the scion. They have found that this makes a much stronger tree and that it produces fewer suckers. It is important that none of the scion of a dwarf tree be planted below ground level, of course, since any roots

For root grafting, cut roots three to four inches in length and about a quarter-inch in diameter.

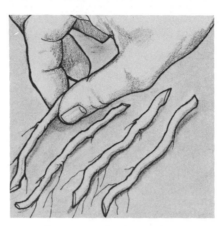

Keep the top ends, the ends nearest the tree, up so they will be planted in correct direction during grafting.

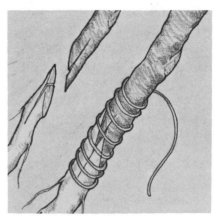

The whip method is usually used in root grafting. The union is held by string and covered with wax.

that grow from the part above the graft will cause it to grow into a full-sized tree.

Root grafts are used by large nurseries to start a great many plants rapidly with a minimum amount of plant material, and sometimes to keep their employees busy during an otherwise quiet season. Because of the skill required to make the grafts and the special care necessary in handling them later, this method is not often used by home gardeners, however.

STEM GRAFT

You would probably use a stem graft on the little tree that popped up in your flower bed. A scion may be placed near ground level, or anywhere on the stem or trunk up to the first branch.

Placing it near the ground allows less surface for suckers to emerge below the graft, but the proximity to the earth makes it difficult to do the grafting, and sometimes encourages collar rot.

This method is used on both dormant and full-growing plants. The scion is inserted into the side of the stock, and the stock can be cut off above the point of insertion.

TOP OR CROWN GRAFT

A scion may be placed on the upper part of the tree to change the variety of the tree, to add a different variety on a larger tree for pollination purposes, or to graft several different varieties on one tree.

Because the tree needs a certain amount of leaf surface to survive, you cannot cut it back too heavily all at once, so you should only graft a few new limbs each year. You can remake a large tree in this way if you want, but it will take years.

We used to have a large tree in our backyard with about eight different kinds of apples growing on it. I tried to add more varieties, but without much luck. Each spring, I grafted on two or three new kinds, but during the winter pruning season, I couldn't remember where the various kinds were, so I inadvertently snipped off a few of the older grafts. If you attempt to create a "table d'hote" you may want to add some identifying marks, so you won't have the same trouble.

CLEFT GRAFT

This is the most common type of graft used by home gardeners. To make a cleft graft, first cut off the branch, or entire tree, at the spot where you intend to place the graft. The scion can be placed close to the ground on a small tree (stem graft), or on an individual branch on a larger tree (top graft), and it is even used occasionally in root grafting as well. It is used mostly on deciduous trees, vines, and shrubs.

When you cut off the branch or tree where you wish to place the graft, with hand clippers, make the cut straight across, at right angles to the branch or limb. Trees less than six feet tall and from one-third to three-quarters of an inch in diameter are the best for grafting. They are vigorous, full of sap, and the cambium layers are easy to line up. If you are grafting onto a larger-sized tree or branch, use a pruning saw rather than clippers to make the cut, because it doesn't crush the wood so badly.

With a knife or grafting tool, split the cut end of the branch, making a vertical cut or cleft through the middle, straight across, about one and a half inches deep. Be careful not to let the tool get away from you and split too far down the stem.

Next, from the scion stick, cut a piece three to five inches long that contains not more than one or two buds. Ideally, the scion should be exactly the same size as the limb or trunk it will be grafted onto, but that seldom happens. It can be smaller, but should never be any larger.

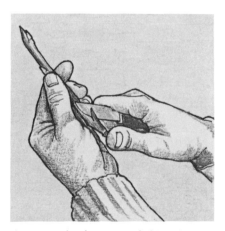

When cleft grafting, make the cut straight across the branch or limb.

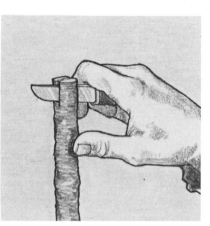

Split the cut end, making a vertical cut one and one-half inches deep.

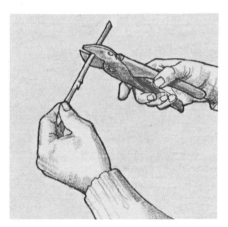

Cut a piece three to five inches long that has one or two buds.

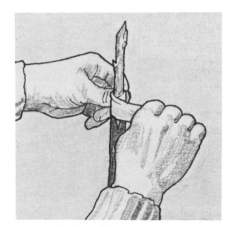

Sharpen the bottom of this scion piece, making a thin wedge.

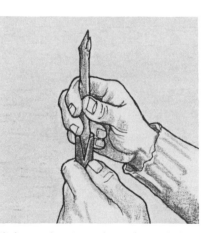

Slide wedge into the split end, line up the cambium layers on one side.

Cover the grafting wound with either rubber tape or freezer tape.

The scion must be inserted in the tree right side up. The part nearest the tree when it was cut is the bottom. The buds should point up.

Sharpen the *bottom* piece of your scion into a thin, smooth wedge by making a sloping cut on each side of the base. Then, with your knife, open the split you've just made in the stock (see drawings), and carefully slide in the scion wedge. Line up the cambium layer (the green layer beneath the bark) of the scion exactly with that of the stock on one side. If the bark on the scion is thinner than the bark on the stock plant, you must set the scion slightly in from the outside surface, to get the proper alignment. Don't worry if part of the cut edge of the scion sticks up above the stock. You can cover it with wax.

Never let the scions dry out. Work swiftly, and make the cuts on rootstocks and scions only when you are ready to do the graft.

Be sure the inside of the scion wedge is no thicker than the outside at the spot where the two cambium layers are to line up, or the stock will not close tightly upon the cambiums and make the necessary close contact. It is best to make the scion wedge a tiny bit thicker on the outside. A close contact is essential so the sap can run freely between them, and there must be no air space between the two pieces of wood.

On larger stocks, two grafts may be inserted, if you wish, one on each side of the cleft. This procedure doubles your changes for success. When both survive, if the graft has been made on a branch, you have two new limbs. If the grafts have been made on the trunk, however, because you don't want a double trunk, allow them to grow for a year, and the following spring cut off the weaker of the two.

Once the scion is firmly in place, cover the entire wound to keep it from drying out. (See Grafting Operation.) Cover all the exposed pieces of cut wood, and fill up any of the crack left where the scion does not fill the split in the stock.

As I mentioned before, there is a wide choice of materials to cover the grafting wound. You may want to use either rubber electrical tape or freezer tape for your first experiment because they are far less messy than wax or tree dressing, and most beginners have good luck with either.

Be patient with the grafts. They may take several weeks to start to grow, and even after that, they are likely to trail behind the growth of ungrafted trees.

SADDLE GRAFT

This method is the reverse of the cleft graft. The rootstock is cut into a wedge shape, and a corresponding split is made on the bottom of the scion. The scion is then placed over the rootstock as shown, and waxed. Rhododendron seedlings are often grafted to named varieties in this way.

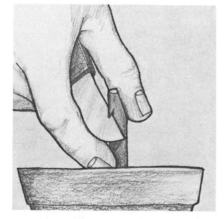

In making the saddle graft, cut the rootstock into a thin wedge shape.

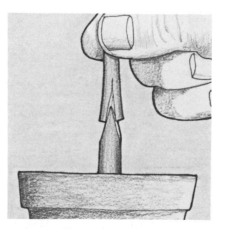

Make a split in the bottom of the scion. Put it over the rootstock.

Apply a layer of grafting wax with a small paint brush.

BARK GRAFT

Some people prefer to make this type of graft, since it exposes more of the cut surface of the cambium layers between the scion and the stock, and can therefore be done with slightly less precision than cleft grafting. The method is similar to that of making a cleft graft, and is used mostly for grafting young fruit, ornamental, and shade trees. The scion should be the same size, and collected in the same way.

A tree or branch from one half to two inches in diameter is usually chosen for this type of graft. Cut the top or branch from the rootstock tree, as if you were making a cleft graft, but do not split the stock. On one side of the stock make a cut through the bark, downward, about two inches long. Shape the bottom end of the scion as shown in the drawing. Note that it is different from the wedge form used in cleft grafting.

Pull the bark from the tree slightly and slide the sharpened part of the scion between the bark and the wood. (See illustration.) Drive one or two small nails through the bark on each side of the slit to hold it tight, and cover all the exposed wood and cracks with grafting wax. If the stock is large enough in diameter, two or three scions may be inserted around the outside.

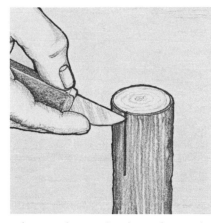

When making a bark graft, cut the top or a branch from the root-stock tree. But don't split this stock.

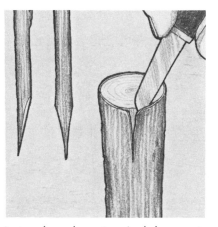

Instead, make a two-inch long cut down through the bark, then shape scion as shown in illustration.

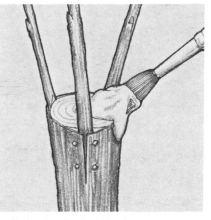

Pull the bark slightly from the tree, and slide scion into position. Hold it with nails; cover with wax.

VENEER OR INLAY GRAFT

Although a veneer graft can be used on many kinds of plants, it is most commonly used for grafting potted evergreens, both conifers and broadleafs, inside a greenhouse in late winter or early spring. In recent years many evergreen varieties have been developed. Some have superior color, such as the Koster Blue Spruce, and others have dwarf, upright, weeping, or other fancy forms.

Seedling plants for stock are potted in the spring, allowed to grow for a season, and moved into a cool greenhouse in the fall. Although they are dormant, they should never be allowed to completely dry out, so water them whenever necessary. The scions should be cut from new growth toward the ends of the branches. They are best gathered on a mild day in late winter when the wood is not frozen, and used immediately.

Unlike a cleft or bark graft, in which a portion of the stock is removed to provide a spot to place the graft, in a veneer graft the top of the stock is retained. The new graft is inserted in the side, where a patch of bark with a bit of wood attached has been removed.

Remove any leaves, needles, or branches from

In veneer grafting, remove branches from the bottom six inches of the plant.

Near bottom of stock, make a notch by cutting down 1¼ inches, through bark and barely into wood.

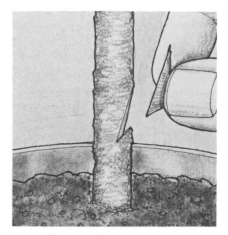

Make a second cut just above the bottom of the first one. Connect with vertical cut. Remove the sliver.

Shape the base of the scion so that it fits into stock opening, with cambium layers touching all around.

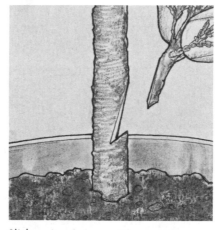

Slide scion into position, resting on the lower notch. Sap must pass freely between stock and scion.

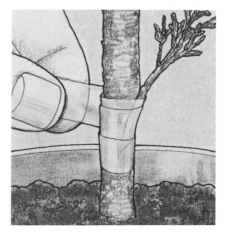

Tie small scions in place with waxed strings or rubber budding strips, or, as here, polyethylene strips.

Because this graft is very fragile, plant should be in a closed propagating case for one or two weeks.

Open case for a few minutes at the end of that time, increasing time until cover is left open all the time.

When healing is complete, remove the top of the old stock. The graft then becomes the new plant or tree.

the bottom six inches of the rootstock plant. Then, with a sliding movement of the knife, make a notch about two or three inches from the bottom of the stock, by cutting downward about an inch and a quarter, though the bark and into the wood about an eighth of an inch. (See drawing.) Without removing any wood, take out the knife, and make another cut starting just above the bottom end of the cut already made. Cut inward, and slightly downward so as to connect with the vertical cut. Then remove the sliver of bark and wood.

The scion should be fairly short, only an inch or two long, unless an especially vigorous species is being grafted. A heavy scion would put too much strain on the stock plant. Remove any leaves or needles from the lower third of the scion.

With a sharp knife, shape the base of the scion to fit exactly into the opening on the stock, with cambium layers touching on all sides. Then slide it into place so it rests on the lower notch. As with all grafts, there must be free passage of sap between the stock plant and the graft, and no dirt or other foreign matter should get in to block this passage or infect the wound. Cover all the wounded areas with grafting wax or some other sealing material.

On a large graft, small nails are sometimes used to hold the union in place. Small scions are tied with waxed strings, rubber budding strips, or narrow strips of polyethylene. Because this type of graft tends to be extremely fragile, the newly grafted plants are usually put into a closed propagating case made of glass or transparent plastic, as described previously. They should be in light, but not full sun, and watered regularly.

Harden Plants

After a week or two, the case should be opened for a few minutes each day to harden the plants, and this hardening period should be increased a little each day. After a month or so, the cover may be left off entirely.

When the healing is complete, remove the top of the old stock either all at once, if the plant is small, or in two or three stages if it is larger. The graft then becomes your new tree or plant.

SPLICE GRAFT

Splice grafting is often done in greenhouses in late winter or early spring on plants with soft pithy centers such as cacti. It is similar to the whip graft except it is done on non-woody plants. First, make sure that both parts to be grafted are the same size. Then cut off the top of the stock plant by making a long diagonal cut, as shown, and a matching cut of the same length and angle at the base of the scion. Cut the head off a pin with a pair of pliers, and insert it half-way down into the cut stock. Then press the scion onto the pin, making sure it fits the stock.

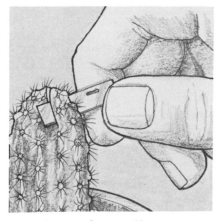

For splice graft, cut off section of the top of the stock plant.

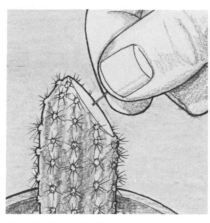

Cut the head off a pin, then insert it down into the cut stock.

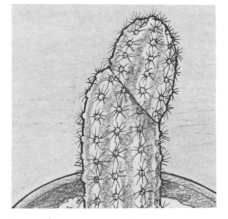

Press the scion onto the pin, making sure it fits the stock.

WHIP GRAFT

The whip graft is often called a whip and tongue graft, because the two parts are cut in such a way that they are firmly locked to each other. (See drawings.)

For this type of graft to be successful, the scion and the rootstock should be as equal in diameter as possible to result in a large area of cambium togetherness, so the graft starts growing quickly.

Both the scion and stock are cut diagonally with a long straight cut. Next, a vertical cut is made in both. The two parts are fitted together, with their cambium layers touching. The graft is then wrapped with nursery tape or waxed string.

A Favorite with Amateurs

Whip grafts can be made on small trees of nearly every kind, as well as shrubs, roses, grapes, and other vines. Fruit trees, especially, were started by the millions in both the United States and Europe by whip grafting, before budding became popular. Because of the high success rate, the method is still a favorite with amateur propagators even though the cutting must be done with careful precision. Like most activities, proficiency comes with practice, and a skilled propagator can make many hundreds of grafts in a day.

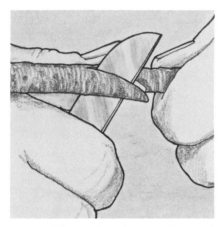

For a whip graft, select a scion and rootstock of same diameter. Cut diagonally with long, straight cut.

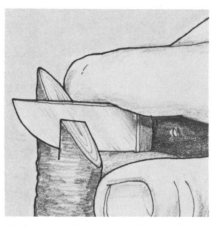

Make a vertical cut on both, then test to make certain they will fit with much of cambium layers touching.

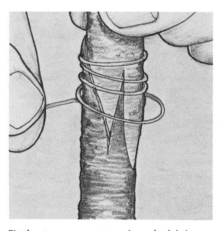

Fit the two parts together, hold them, then wrap the graft, using either nursery tape or waxed string.

AFTER-CARE OF GRAFTS

I lost a lot of grafts during my beginning years of grafting because I was too impatient. Grafts take longer to start growing than do the branches already established, and can sometimes look pretty dead among all the green sprouts. They often start to grow three or more weeks later, so don't give up too soon. In the meantime, don't touch the graft or allow anything to knock it out of alignment. Even after it starts to grow, a new graft is still very fragile, and the wind or a slight touch can snap if off easily. Because all the strength of the root system is supplying a small scion, the growth is often very rapid once it gets going, however, and this makes the fragile union even more brittle. On trees grafted close to the ground, it is good insurance to put a small stake nearby and tie the graft to it as soon as it has grown tall enough. Trees grafted higher, on the branches, are harder to protect.

Remove Side Sprouts

Pinch off all the side sprouts that form the first summer on grafted fruit and shade trees, and encourage the new graft to grow straight with a

On trees grafted close to the ground, put a stake in the ground and tie the graft to it to protect it.

Pinch off all side shoots that form the first summer on grafted trees, to encourage graft to grow straight.

Pinch or snip off any sucker shoots that sprout beneath the graft, so they don't crowd out the grafted top.

single stem. On lilacs, roses, and similar shrubs, the top sprout may need to be pinched off to encourage side branching, so the plant will have a bushy shape. On all grafts, the sucker shoots that sprout from below the graft should be pinched or snipped off. Otherwise, because they are so vigorous, they will quickly crowd out the grafted top.

If your grafts are in pots, the nutrients are likely to leach away rapidly because of frequent waterings, so some liquid feeding or other fertilizing during the growing season will be necessary for good growth. If the plants are in the ground, however, and the growth of the new grafts appears good, it is not a good idea to feed the plants at all the first year. Top growth that has grown too fast is not only very brittle, but in the north it can fail to harden before the first frost, and be injured by sub-freezing temperatures.

BUD GRAFTING

Although bud grafting, or budding, as it is commonly called, is technically a form of grafting, many propagators refuse to admit it. I once heard a nurseryman say with disdain, "Our trees are budded, not grafted," as if budding was in some way a superior technique. No matter how it is described, budding, like grafting, is the joining of two plants to form a new plant. Instead of attaching a scion to an understock as in regular grafting, however, only a tiny dormant bud is placed on the stock. Over the years I have bud-grafted thousands of trees, but I am still amazed at how a whole new tree grows so quickly from a small bud, and how such a tiny object can change one variety of tree to another.

Quick and Easy

Many propagators prefer budding to grafting because it is faster, easier, and less messy. No wax is necessary, and cambium layers need not be aligned. Although plants can be bud-grafted in early spring with a dormant bud, it is usually done in the summer with a newly developed, latent bud which is taken from under the stem of a live leaf. Because the mid-summer season in which budding may be done lasts for several weeks, there is not as much pressure to get the job done in a hurry as in scion grafting in the spring. The time sequence also gives you a second chance in the same year to re-graft any spring attempt that may have failed.

Although many professional and amateur propagators prefer budding to grafting, most agree that in the plant nursery there is a place for both. The side grafting of evergreens, splice grafting of cacti, and bench grafting of fruit trees cannot be satisfactorily duplicated by any kind of budding. Many plants, such as nut trees and certain shade trees, have a better success rate when propagated by regular grafting, but most of the millions of roses and fruit trees that are sold each year are propagated by budding.

NATURAL GRAFTING

Not all grafting is done by humans.

The plants themselves often achieve this, usually after branches or roots have been pressed closely against each other for a long period.

A common natural graft often seen is on climbing ivy. Here the graft is so efficient that sap readily passes from one plant to another. This is demonstrated when one plant is cut off at its base, but continues to grow through its link with the root system of another plant.

These grafts are created when two branches are in close contact over a period of years, a contact so close that the bark at the point of contact never thickens, and eventually the graft is achieved.

Studies have determined that these natural grafts are achieved in the younger parts of the plant, when the bark is still smooth and thin.

They are far more common, too, in plants that have smooth barks; thus they are relatively common in beech and maple trees.

EQUIPMENT

A pair of hand clippers is necessary to gather the budwood. A sharp knife is essential, too, and if you work with many buds you should keep a sharpening stone handy. You'll also need some kind of material to hold the bud in place. Yarn, raffia, and string were all used in earlier times, but were unsatisfactory because, unless they were cut away shortly after the operation, they would girdle the tree as it grew. I have tried electrical tape, Parafilm, and ordinary transparent tape, but none of them worked well.

I have had the best results with rubber budding strips when budding fruit and shade trees, and with plastic patches, when budding roses. Both are available from garden and nursery supply stores (see Appendix). You may make your own strips, if you prefer, by cutting narrow pieces from a wrap such as Saran, or a plastic bag. It is imperative that whatever material you use is capable of expanding slightly, as the tree increases in diameter during the latter part of the growing season.

MIDSUMMER BUDDING PROCEDURE FOR TREES

Summer budding, or fall budding, as it is sometimes called, is the easiest and most popular form of bud grafting. Spring budding and June budding are also possible, and are discussed later in this chapter.

Conditions are right for summer budding when there is sap under the bark, and a fat, well-developed bud under each new leaf along the new growth of the tree. Sap is usually present from the time growth starts in the spring until the tree stops growing in late summer, but the bud is ready only after the new wood has started to harden slightly. In the northern states the budding period is usually from late July through late August, but it can be done over a longer period in more southern regions. If budding is done too early, the new bud isn't large enough to survive being transplanted. If it is attempted too late in the season, the bark doesn't "slip," or separate, from the wood easily, and it is difficult to successfully insert the bud. During a dry growing season, budding may also be tricky because less sap is present, and irrigation may be necessary for a few days ahead of the operation.

Although the season is long, it is a good idea to begin budding as early as possible so a new bud may be put in the understock if the first one fails.

The Budwood

Select for budwood only healthy branches that have grown the same season, and preferably those that have large buds. The length of the new growth on a tree varies anywhere from a few inches to two feet or more. I find that branches eight to fifteen inches long are convenient to handle.

As a rule, young trees make a large amount of growth each season and it is easy to find good budwood, but very old trees grow so little that it may be a lot tougher to locate good buds. Even on young trees, the growth may be small in dry years or following a hard winter.

As with regular grafting, discard the wood at each end of the branch and use only the fat buds in the middle. The smaller ones at the top end tend to be immature, and those at the bottom end are even more latent, and less inclined to grow. Also, as with grafting, be sure to use a leaf bud, and not a flower bud. You can easily separate the narrow leaf buds from the much fatter flower buds. The new growth of most trees and shrubs is not likely to contain flower buds.

Collect the budwood at least a day before using it. Although it can be held for three or four days if it is kept cool and not allowed to dry out, I like to gather the budding sticks the day before. Put them in a pail with their bottom ends in water and take them to a shady spot. Pinch off the leaves, leaving about a quarter-inch of stem above each bud to serve as a handle later, when you insert the bud. Be sure to label each branch, or bundle, if you are working with several kinds, and store them with their cut ends in water in a cool place such as a basement. Keep them in

A least a day before they're needed, cut budwood from healthy branches.

Pinch off all leaves, but leave a quarter-inch of stem as a handle.

Label each one, and store them in cool place with cut ends in water.

water until you are ready to cut off the buds, just before inserting them into the stock.

In large commercial operations, usually one person does the actual budding and one or more follow him to do the tying on. Several thousand buds can be inserted each day in this manner, and large nurseries like the way so many trees can be started with a minimum of propagation material. Your budding operation, like mine, is probably considerably smaller, but even so, you will no doubt be surprised to find how easy and smooth the operation goes, after you have done a few.

The Understock

As in regular grafting, the understock should be the same variety as the plant from which the bud was taken, or very closely related. The same compatability problems exist, so read the grafting chapter closely before choosing an understock. Try to avoid seedlings from unhealthy trees or those that are not vigorous.

Although it is possible to bud graft small top limbs on a large tree, the procedure is used ordinarily to convert young sapling trees, six feet tall or less, to a desired variety.

Because budding does not work well if the bark is too thick, it is difficult to place a bud on a tree or branch that is much over a half-inch in diameter. Usually budding is done on trees or limbs from three-eighths to one-half inch in diameter. A tree two or three years old, with one single stem that has leaves nearly to the ground, is more likely to be full of sap and have bark that slips nicely.

Placement Of The Buds

As in scion grafting, you have a choice of places to put the bud. On a young tree, I place it as close to the ground as it is convenient to work, usually about two or three inches above the soil. Some northern horticulturists, however, prefer to bud their trees nearly a foot above the ground, believing that, because it exposes more hardy rootstock, the new tree will be more hardy.

If, instead of budding on the stem of a small tree, you wish to insert a bud in a small branch of a larger tree, look the branch over carefully first, and place it at a spot where you most want the new limb to grow. The following year you will be cutting off the part of the limb that extends beyond your bud, so choose a place where the new limb can grow to a good size without interference. Don't try to bud too large a branch. Just as in budding on the stem, it is difficult to place the bud on a limb larger than three-quarters of an inch because the bark will be too thick.

Sanitation

When bud grafting it is especially important to keep everything clean and not let anything unsanitary touch either the bud or the cut you have made in the bark of the stock. If you accidentally drop the bud on the ground, throw it away and cut another. Don't put the bud in your mouth or in dirty water before inserting it, either. It is not necessary to sterilize your knife or put the budwood in sterile water, but try to keep all foreign matter from the cut.

THE T-BUD

This is the most common budding method, and it is, I feel, one of the most satisfactory ways for a beginner to graft a tree.

(1) Cut a T shape through the bark of the tree to be budded, making sure that the cut does not penetrate the wood. Generally, a horizontal mark a half-inch long and a vertical one, three-quarters of an inch long, is about right. The size of the cut is determined, however, by the diameter of the tree being budded and the size of the bud being inserted. Don't try to put a large fat bud on a tree that is too small for it. The bud must fit inside the bark in order for the bud and stock to grow together successfully. Many prop-

agators like to put the bud on the north side of the tree, feeling that it is less likely to dry out when shaded from the sun.

(2) Open slightly the bark flaps at the sides of the T, and check to see that the area is moist. If the bark doesn't separate from the wood easily, there will not be enough sap to support the new bud after it is inserted. If you have trouble opening the bark, or it breaks into pieces, forget that spot and try to find another, either on the same tree, if it is large enough, or on another one.

(3) Insert a knife approximately one-third inch below the base of a bud on the budwood stick and, with a sliding upward motion, cut out the

T-bud is a fine one for beginners. Cut a T shape through bark of tree; careful, don't cut into the wood.

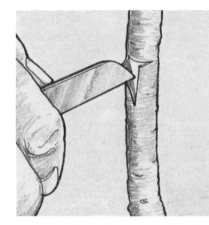

Open the bark flaps at sides of T. Bark should separate from wood easily, indicating sap is there.

Insert knife below base of the bud, and carefully cut out the bud, and with it, a small sliver of wood.

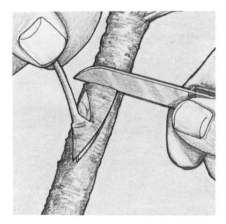

Make a horizontal cut above the bud, so that you can cut it from budstick.

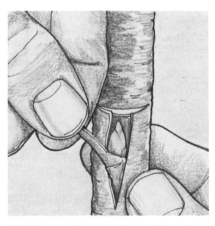

Slip the bud, top-side up, into the opening, and secure it against the sap-covered surface of the wood.

Tie the bud securely within the bark, using a rubber budding strip that is wrapped above and below the bud.

bud, including a small sliver of wood. Make a horizontal cut just above the bud, to sever it from the budstick. The piece should be the shape of a tiny shield. Hold it by the leaf stem, which makes a convenient handle.

Some propagators feel uncomfortable about putting even a thin sliver of wood into the T, feeling it makes it more difficult for the union to heal. They prefer to cut off a slightly thicker piece of wood initially, as they remove the bud, but they dig out all the wood with their knife before actually inserting the bud and bark into the stock. They feel that this procedure permits a much tighter union. If you try this method, be careful not to tear the bark around the bud while removing the wood.

(4) With your knife, open the bark flaps on the rootstock tree again. Slip the bud, top-side up, into the opening, and secure it tightly against the sap-covered moist surface of the wood. Make sure the top of the shield doesn't stick up above the horizontal cut of the T.

(5) Tie in the bud, or in some other way secure it within the bark. It is important that the bark be held tightly, because if it pops open, the bud will dry out—the most common cause of failure in budding. If you use a rubber budding strip, begin to wrap it around the stock slightly below the bottom of the T, and stretch the band somewhat

so it will be tight. Wrap, as if you are winding an ace bandage, up to the newly inserted bud and all around it, but don't cover the bud or the leaf stem you have used as a handle. Secure the end of the rubber strip by stretching out a piece of that which has already been wrapped, and tucking the end under. Don't remove the stem handle yet.

If you use a clear plastic strip to secure the bud, you may break off the leaf stem handle and wrap the area in the same way, except that you should cover the bud itself with one, and only one, layer of plastic. Be sure the wrapping is tight and that air is completely excluded. (See "Budding After-Care" for procedure following budding operation.)

Try Only One Bud

Sometimes gardeners wonder if it might be better to put two or three buds on a tree, rather than only one, hoping that with the law of averages, at least one might grow. On a small tree or branch, the budding operation wounds the trunk and interrupts the flow of sap considerably, so it is better for the tree if you do only one at a time. Usually there are very few ideal spots on a small tree for budding anyway, and it is better to use only one, hope it will work, and save the others in case you need them later.

INVERTED T-BUDDING

Inverted T-budding is done in the same manner as regular T-budding, except that the cut is made upside down. The bud is inserted upright, however, just as in regular budding. Propagators often prefer this method to regular T-budding because during a wet season, rain is less likely to collect around an inverted T-bud and make it rot. It is also used for trees that bleed badly, such as chestnuts. Some horticulturists feel that inverted buds grow into stronger, better-shaped trees, and that fruit trees budded in the manner bear at a younger age. I have tried both methods and have not seen a great deal of difference in the results, although I find regular budding much easier and faster.

Inverted T-budding is good to try during rainy season.

CHIP BUDDING

Chip budding is more difficult than T-budding because it is more exacting. Unless the parts fit together closely, they will not unite. This method is also more time-consuming than either T-budding method. Usually it is done professionally only when the bark does not slip easily enough for regular T-budding.

With care it may be done successfully early in the spring or late in the summer, and nurserymen sometimes resort to it even in mid-summer, if a drought has caused a shortage of sap in the wood.

(1) Cut a V-shaped piece of bark and wood completely out of the stock as shown in the drawing. The resulting notch should be the same size as the wood and bud you are planning to insert.

(2) Cut a bud from the budstick, including a chip of wood about twice the size of that used for a T-bud. Cut it in a shape and size as exactly like the one in the stock as possible, since they must fit together tightly. If you are doing the budding in the summer, remove both the leaf and the leaf stem.

(3) Place the bud chip in the cut-out area of the stock and hold it tightly. Since there are no bark flaps to secure the bud in place in this kind of budding, you must fasten it in some way other than by budding strips. Nursery grafting tape is often used for this purpose, but you can get good results by using a small piece of plastic food wrap stretched tightly. Secure it in place by tape or by tying it tightly with rubber strips. If you use tape, leave the bud itself exposed, but if you use clear plastic, cover it with one layer. Remove the tape or plastic about three or four weeks later.

As in T-budding, the top of the plant is not cut off until the bud has united with the stock. In summer and fall grafting, the cutting is done the following spring. In spring grafting, however, it is done about two weeks after budding, as long as the bud appears to be alive. Unwrap the bud at the same time.

Cut a V-shaped piece of wood out of stock. Size must match insert.

Cut a bud from the budstick, making it same size as hole in the stock.

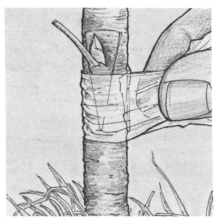

Fit the two together and hold in place with grafting tape.

OTHER TYPES OF BUD GRAFTS

Several other budding techniques are used by commercial propagators in special circumstances, or for certain types of trees. Most are not easy for home gardeners because the operation is complicated and extremely exacting.

Patch budding is often used on trees that have thick bark, such as walnuts and pecans. A piece of bark with the bud on it, but no wood, is removed from the scion branch and inserted on the rootstock where a patch of corresponding size has

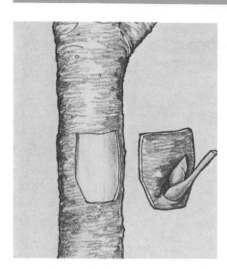

A patch bud is used for trees having thick bark. The patch and opening must be the same size.

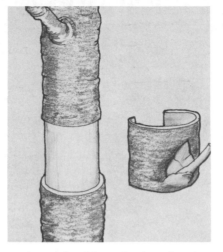

With ring budding, bud and a ring of bark are fitted where a ring has been taken from the stock.

I-bud is put into position like the T-bud, but through a cut shaped like an I. Tape bud into position.

been removed. Both the patch and the opening must be shaped exactly the same, so the fit will be perfect.

The area, except for the bud, is then covered with nursery adhesive or grafting tape, or the patch is tied in with waxed grafting thread, again without covering the bud.

Ring budding is similar to patch budding, except that a ring of bark is removed from completely around the budstick, and inserted where a ring of the same width has been removed from the stock tree. The bud is secured and protected in the same manner as a patch bud.

I-budding is similar to T-budding, except that cuts are made like a capital I. A bud with a rectangular-shaped piece of bark is used, and, as with the regular patch bud, no wood is left with it. The patch is taped in, as described previously. Patch budding is also useful on trees that have bark that is too thick for regular T-budding.

A Helpful Tool

Special knives with double blades are sold by nursery supply houses to make it easier to meet the exacting requirements of patch and ring buds.

BUDDING AFTER-CARE

After a week, check a few of your newly inserted buds to see if they are viable. Carefully break off the little leaf stem handle. The bud will stay dormant until the following spring, but if it is still green and lively looking at the point where the stem was removed, it is receiving sap from the stock, and you probably have a "take."

If not, there may still be time to insert another bud somewhere else on the tree, or on another tree.

If you have used clear plastic wrap, it is usually possible to see the bud and tell whether or not it is still green. Of course you may unwrap, examine, and rewrap it, but it is safer for the bud to leave it alone as much as possible.

If you insert another bud, always use fresh budwood. Don't try to save budwood for more than four or five days, or the results will be disappointing.

A month or so after inserting the bud, check it again. If you have used rubber budding strips, they should be starting to rot and fall off. Their disappearance will give you a good look at the bud, which should be fat and healthy-looking. If you have used tape or plastic, carefully cut it away, so it won't smother the bud. No wrapping material should be left on over the winter. The bud, like the other dormant buds on the plant, will just "sit there" until spring, without growing, but it shouldn't be constricted by anything.

If you have trouble with mice and rabbits, be sure to protect your newly budded trees in some way. I've found that the little creatures like nothing better than chewing up young trees, especially if they have potential value.

Fruit And Shade Trees

Early the next spring, when the tree is still dormant, cut off the top of each tree by making a sloping cut away from the bud, as shown in the drawing. It should be about a half-inch above the spot where you inserted the bud. Don't be too upset if the bud no longer looks lively, because it is likely to have turned a dull brown over the winter and appear dead. Have patience and resist the temptation to pick at it to see if it is viable.

Citrus plants are often not cut off completely at the end of the dormant period, but instead, the top is partially cut off and bent over. The top foliage feeds the plant, but the "breaking over" forces growth into the bud. Later, as soon as the bud has grown a few inches, the top of the old plant is completely removed.

Some propagators heap piles of sand or sandy soil up over the buds for the first winter to pro-

tect them. The sand is moved away as soon as weather warms in the spring.

Check the plant frequently during the growing season to see that other sprouts are not starting to grow below the bud. If they appear, rub them off as soon as you spot them. When your bud has grown a few inches, put in a small stake, and tie the new little tree to it with a plant tie or yarn, so a hard wind won't snap it off. It will grow with a slight bend the first year or two, but will soon straighten out. By pinching and snipping, encourage the tree to grow with a tall, straight stem, and discourage any side branching until it is three or four feet tall.

Do not over-fertilize a newly budded tree, but allow it to grow naturally. Too-fast growth will break easily, and often fail to harden before winter.

Shrubs And Roses

Shrubs and roses are treated in much the same way as trees, but instead of a single main trunk, they should be made to grow into a bushy shape with numerous branches near the ground. To accomplish this, place the bud as low as possible. When it starts to grow, the top sprouts should be pinched back occasionally to force more side branching.

After the new plant has grown for a year, transplant it once again the following spring so it can be set a few inches deeper. Deep planting helps prevent excessive suckering and encourages low branching.

During early spring after budding, cut off the top of each tree.

Check frequently for sprouts that grow below the bud. Rub them off.

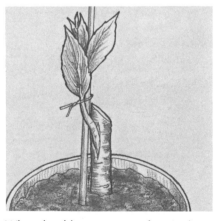

When bud has grown a few inches, put stake nearby, tie tree to it.

OTHER TIMES FOR BUDDING

Although summer is the most common time for budding, for one reason or another you might like to do it at some other time.

Spring

Spring budding is done during the regular grafting season, but instead of joining a scion to the stock, you insert a dormant bud that developed the previous summer. It should be done as soon as there is enough sap in the stock so the bark will slip from the wood easily. The buds may be gathered nearly anytime in late winter until the time growth starts, and kept in a refrigerator. It is necessary, just as with regular spring grafting, that the buds be still dormant when they are inserted; but the understock should be full of sap, and the buds on it should be just starting to show green tips.

The process of bud grafting is the same as in summer, and T-budding, inverted T-budding, or chip budding may be done. Since the bud will begin to grow the same season, however, about two weeks after the operation you should cut off the top of the tree at an angle, a quarter-inch above the newly inserted bud. If the bud is on a limb, remove the tip of the branch beyond it, also with a slanted cut. This will force the new bud to grow, and from it will come your new shrub, tree, or branch.

Spring budding has never been very popular because regular scion grafts are usually made at that time. Also, spring buds usually take quite a while to start growing and do not make as much growth the first year as spring grafts or as buds that were inserted the previous summer.

June

June budding is common only in southern areas where a long growing season makes it feasible. The new buds form early and as soon as they are large enough they are removed in the regular way, and inserted in the stock, either by chip budding, regular budding, or inverted T-budding. A week later, the top of the tree is cut back to stimulate the growth of the bud. Since the tree is in a fast-growing condition, the entire top should not be removed all at once, however, but in two stages. A week after budding, the top is cut off a few inches above the new bud, leaving a few leaves and perhaps a branch still on the tree. These are necessary to manufacture food for the tree until the new bud has made sufficient growth to be able to do it. As soon as the new bud has grown a few inches, the remainder of the old top is cut off at an angle, just above the new sprout.

June budding is used mostly for propagating plum, apricot, almond, peach, and other fast-growing trees.

In regions where the climate permits, it is possible to grow a tree from budding to saleable size in the same season.

TISSUE CULTURE

Plant propagation by tissue culture is another example of science fiction taking place today. It requires no stools, grafts, seed beds, or mist houses. Instead, plants are cloned in test tubes in a sterile laboratory.

Every tiny cell in each plant has the awesome potential to grow into a plant exactly like its parent, if given the right conditions. Any cell from a root, leaf, or stem has this capability, but tissue from within a bud is most often used, since that portion is actively growing, and responds quickly in the laboratory.

Once started, the culture can be increased indefinitely, or divided and grown into plants. A portion may also be held dormant for future use —a plant bank to be drawn upon when needed.

The possibilities for tissue culture are mind-boggling. From one tiny piece of bud or stem, an incredible number of plants can be produced in a small space, within a very short time. Up to a million plants can be started in a space twenty by twenty-five feet in one year.

If you have bought trees or plants recently, you may already own some that were started by this method. Although it has been feasible for only a few years, thousands of strawberries, orchids, ornamentals, perennials, blueberries, and dwarfing rootstocks for fruit trees are already being produced by large nurseries. Each year there are increasingly large numbers of shade trees, evergreens, roses, and interior landscaping plants being propagated by tissue culture.

Like most "new" developments, this method of propagation has its roots in botanical experi-

ments that began many years ago. In 1934, P. P. White found a way to grow tomato roots continuously in laboratory jars by feeding them yeast extracts, thus supplying their B vitamin needs. Later, in 1939, White in the United States and two French scientists, Gautheret and Nobecourt, working independently, learned how to culture plant tissue indefinitely in the laboratory. A major breakthrough came in the seventies, when Dr. Tsai-Ying Cheng of the Oregon Graduate Research Center in Beaverton, Oregon, developed a process that made tissue culture commercially feasible.

Advantages

In addition to the speed of production, there are many other advantages to increasing plants by tissue culture:

(1) Certified, disease-free plants such as strawberries and geraniums can be easily produced in large numbers and sold uncontaminated. This characteristic of tissue-cultured plants is most worthwhile, because much of the plant life currently being propagated is loaded with disease, which is also propagated. Dozens of different kinds of plant viruses have been identified in everything from fruit trees to houseplants, and plant pathologists agree that these diseases weaken the plant, decrease the production of flowers and fruit, and shorten its life. Disease-free plants can be of great benefit to all of agriculture by increasing productivity, and requiring the use of far fewer chemicals.

(2) The price of new plants should remain reasonable far into the future because of the

speed with which they can be produced. Not only is the process much more cost-efficient than propagation by other methods, but it will no longer be necessary to go to the expense of cultivating and replacing large numbers of stock plants that now must be grown and tended to provide the necessary grafts and cuttings.

(3) Trees that are usually grown by grafting can be grown on their own roots by tissue culture, instead of being grafted on some other rootstock. Hence, the problem of possible graft incompatibility is gone, as is the nuisance of suckers that grow from below the graft. Another benefit is that if mice or rabbits chew the bark and kill the top of a favorite tree, the tree may still be saved. Any new sprouts that emerge from below the girdled part will be the same variety as the original tree, rather than from an inferior rootstock sucker, as on a grafted tree.

(4) The new varieties of plants and trees that are constantly being developed can now be made available quickly all over the country. Previously, a decade or more might pass before a new variety could be produced in large enough numbers for public use.

(5) Nurserymen will have better control over their inventory because they will no longer need to guess what the public will want many years in the future. A tree that could be grown to a height of four to six feet by conventional methods in two or three years, for instance, can be produced in as little as five months by tissue culture.

Some Problems

Cloning is not the answer to all propagation problems, however. Grafting is still an essential art for certain kinds of plants. Tea roses, for example, have been so over-hybridized that they require a hardy, vigorous-growing rootstock to make them grow well in most soils; and there is no reliable method of producing many dwarf fruit trees without a graft.

There are other problems with tissue culture. Occasionally mutations or "sports" appear on fruit trees and certain other plants. A tree that produces all red apples, for instance, might one year have a limb that begins to produce striped apples. Although many of our best fruit varieties first appeared in this way, sports have been known to produce inferior variations, as well.

Not only might cells be taken inadvertently from a sport limb, but botanists fear, also, that in the laboratory the speeding up of cell growth may produce mutations at an accelerated rate. The rapid propagation of a quite different plant from the one intended might occur, and laboratory personnel would be completely unaware of it. Experiments with chemicals such as benzyladenine are being conducted in attempts to make such mutations less likely.

Generally, the plants that are easiest to grow from seeds or cuttings are also the easiest to culture in the laboratory. Most herbaceous plants, such as perennials and vegetables, start well and grow rapidly. Woody plants are more difficult. Among other reasons, the tiny portion of the stem or bud used for cloning, called the explant, must be sterilized carefully before it is put in the culture medium, and it is much more difficult to get the explant of a woody plant absolutely sterile.

Exacting Conditions Necessary

The sanitary requirements of tissue culture are very exacting. Unlike most propagation, when you worry only about controlling harmful fungi in soil, in tissue culture the air, water, and all materials used must be absolutely free of yeasts, fungi, and bacteria. Consequently, all materials, air, and water entering the laboratory area must first be sterilized.

Temperature, light, humidity, and timing are carefully controlled, also. Super-sensitive electronic scales are needed to measure the ingredients used in the culture. A dependable source of electrical power is essential to sterilize the equipment and to maintain the proper atmospheric conditions. The equipment involved makes the initial investment costly.

How to Begin

To begin the propagation, the tiny explant is cultured in a mixture of agar (a jelly-like substance made from seaweed); a source of energy, which is often sugar; a growth regulator; and various vitamins. The culture mix varies, because different species of plants do not respond to the same formulas. Also, different varieties of the same species, and sometimes various strains of the same plant variety, require slightly different cultures. Not only does a Bartlett pear, for instance, require a culture that is different from that needed by a Bosc, but a Bartlett pear from one area of the country often requires a different culture than a Bartlett pear from another. The varying responses of plant cells to the culture mix is one reason that detailed instructions for tissue culture are not more widely available. A laboratory must develop the cultural methods for each variety of plant it produces.

The test tubes used for culturing the explants are mounted on a slant, so the exposed surface will be larger, and more room will be available for growth. At regular intervals, the growing mass is divided. Part may be saved for future culture, and part divided into tiny sections with scalpels and tweezers. Each of these minute sections is then placed in a small glass jar containing a different culture. The size of the mass stops increasing at this point, and each section begins to develop roots and sprouts. Soon it starts to look like a small plant.

The pH of the cultures must be carefully controlled throughout the process. To make it more acid, a few drops of dilute hydrochloric acid are added or, to make it more alkaline, sodium or potassium hydroxide is used. For most cultures, a pH of 5.7 is necessary.

Transfer Difficult

As you might imagine, it is even more difficult to transfer a young plant that has been cultured in a laboratory jar to the outside world, than to move one outdoors from a mist house. The transfer must be made carefully, in several stages. First, the plants are moved into a greenhouse where the temperature is high and the humidity is nearly 100 percent. After a period of time, they are either moved into another area with slightly lower temperature and humidity, or left in the same place with changed environmental conditions. Each change acclimates the plants gradually to an outdoor climate.

On a Small Scale

A laboratory, in order to recoup its investment, must produce a huge number of plants, which requires that it operate year-round. A sizeable amount of greenhouse space must be maintained to protect and grow the plants until they can be sold.

Amateur experimenters work on a scale that is

much less grand than that of commercial operators, and their equipment is less sophisticated. For example, many use pressure cookers instead of autoclaves for sterilizing test tubes, and germicidal lamps instead of Laminar Flow Hoods to sterilize the air. They use sterile disposable containers to avoid the need for sterilization, and make use of canning and baby food jars as substitutes for costly laboratory glassware. Most buy prepared culture mixes in small quantities instead of raw materials in bulk. According to some reports, it is possible to get into the tissue culture business in this manner for less than $1,000.

Although at present the process seems feasible only for commercial production and may never be the most practical method for amateur gardeners, culture kits are already available for students and experimenters. Scientific knowledge is not a requirement, and if you can read a cookbook, you can master the fundamentals quickly and easily. Because most university research is highly specialized, small experimenters are more likely to introduce the changes that will make the method more feasible in the future.

Future Impact

There is every indication that tissue culture will impact our future in endless ways. To name a few, it may help us grow superior forest trees that will greatly speed up the production of timber, alcohol, paper, and firewood. Forthcoming, too, should be disease-resistant fruit trees that are compact and need little care, tight-growing Christmas trees and ornamental evergreens requiring almost no shearing, and maple trees with sap that is two or three times sweeter than average. Tissue-cultured plants will be part of our future as surely as are the transistor and silicon chip. For more information about sources of supplies and equipment, and books on the subject, refer to the Appendix.

Although watching a blob of plant tissue turn into a tree in a laboratory jar sounds terribly exciting, most of us will prefer to get closer to nature by watching grafts come to life and seedlings push through the soil in our own backyard. We can be confident that the old familiar methods of propagation will always be with us, not only because they will often prove more practical, but also because they are more fun.

Part II
SPECIFICS
OF PROPAGATION

FRUITS AND NUTS

Because most fruits and berries do not come true from seed, they are reproduced asexually. Small fruits and berries are usually propagated from layers, suckers, or cuttings, and tree fruits and the improved varieties of nuts are started by regular or bud grafts. Commercially, many fruits and berries are now being propagated by tissue culture, but most home gardeners use the more conventional methods. Seeds are used to propagate the rootstocks for tree fruits and nuts, and they are used as well for developing new varieties.

ACTINIDIA CHINENSIS. See GOOSEBERRY, CHINESE

ALMOND *(Prunus amygdalus).* The T-bud is the most common method of propagating the almond. The budding may be done in late spring, early summer, or fall. Neither hardwood, softwood, nor root cuttings have proven very successful.

Almonds can be bud grafted successfully on plum, peach, apricot, or cherry trees, because they are so closely related. Usually almond seedlings are selected for the rootstocks, although peach seedlings are sometimes used in almond orchards where the trees are cultivated and irrigation is used.

ANANAS COMOSUS. See PINE-APPLE

APPLE *(Malus sylvestris).* Most named varieties are propagated by grafting and T-budding, although some of the more vigorous varieties can be reproduced by softwood cuttings and root cuttings. Although sometimes suckers may be dug and transplanted, these should be used for grafting stock only, because they have probably sprouted from the wild root below the graft, and will therefore produce only inferior fruit. Layering is possible if a suitable branch growing above the bud or graft can be found, but it may take two or more years before a good root system develops.

Air layering, too, may be done, although roots may not develop for several years.

Grafting

Commercially, apples were formerly propagated mostly by whip grafting scions of good apples on roots or pieces of roots from seedling apples in basements or barns in midwinter, in a process called bench grafting. Home gardeners usually find cleft grafting, bark grafting, or T-budding on seedlings the most practical method. Both the seed and seedlings are available from nurseries and seed houses. If dwarf trees are wanted, rootstock trees may also be bought, or sometimes it is possible to start a few by carefully digging up the suckers from around the trunk of dwarf trees.

Rootstocks

Since the roots influence the ultimate size of the tree, they are usually chosen to produce the size of tree wanted within the limits of hardiness and adaptability to soil conditions. For standard, full-sized trees, plant the seed of vigorous growing varieties that do well in your area. If not cut back repeatedly, these will ultimately grow to a height of eighteen to twenty-five feet tall, and sometimes more. Commercially, seeds are often obtained from cider mills or fruit processing plants, and seeds from crabapples and flowering crabs are also sometimes used. Wild apple seed is planted occasionally, but is not usually recommended because the rootstock may influence the flavor and ripening of the fruit. McIntosh, Yellow Transparent, and Dolgo Crab seedlings are popular in the North, and Red Delicious, Golden Delicious, Winesap, and Rome, further south. The seeds of Baldwin, Jonathan, Gravenstein, Rhode Island Greening, and Wealthy do not usually give good results.

Because dwarf and semi-dwarf apples are popular, many nurseries propagate rootstocks by layers, stools, and tissue culture. East Malling (EM) and Merton Malling (MM) from England are often used as dwarfing stocks. These rootstocks are available from nurseries listed in the appendix. The rootstocks vary somewhat in vigor and their adaptability to different soil conditions, as well as the ultimate size of the tree produced. They also vary in hardiness. Most of the Mallings and East Mallings are unsuitable for the coldest sections of the country. Dwarfs also tend to be shallow rooted, and may not be the best choice where hard winds are prevalent.

Dwarf Rootstocks

EM 2. Semi-dwarf, ten feet or less. Not good for clay soils.

EM 7. Widely used rootstock. Eight feet. Does well in heavy soils and suckers very little.

EM 9. Very dwarf, about six feet. Good for espaliers, apple hedges, and small areas, but should be staked.

EM 26. About eight feet tall. Needs staking, and often gets winter injury in the North.

EM 27. Most dwarfing of all. Hardier than most Mallings, but needs a well-drained soil.

MM 104. Semi-dwarf. About eleven feet.

MM 106. About nine feet. Needs well-drained soils and long growing season.

MM 109. Semi-dwarf. Productive, non-suckering.

MM 111. Vigorous, productive, and good for warmer climates. This is used widely in commercial plantings.

Other Rootstocks

Alnap 2. A hardy, vigorous-growing rootstock from Sweden that produces a tree about eleven feet tall.

Robusta No 5. A vigorous rootstock from Canada that is best in cold climates and not satisfactory in areas with longer growing sea-

sons. It produces a large, semi-dwarf, twelve to fifteen feet tall. Popular for some time in the Northeast, it is now being replaced by seedling and other roots, because of virus problems.

APRICOT *(Prunus armeniaca)*. Named varieties of apricots are ordinarily propagated by T-budding them on apricot, plum, or peach seedlings. Regular budding is the usual practice, but spring budding can also be done. Apricot trees can be layered or air-layered, but rooting will usually take two or three years. Propagation by softwood cuttings is possible, but not easy.

Seeds may be used to grow rootstock for grafting, or for developing new varieties. The chance of getting a good variety from a seed is somewhat better with apricots than with apples, however. Fall planting or stratification of the seeds is recommended.

Plum seedlings make the best rootstocks if the trees are to be planted on heavy, damp soils; but peach or apricot seedlings are often used when the soil is warm and sandy. Seedlings of American wild plums and of hardy apricots such as Scout and Moonglow should be used as stock where hardiness is important. Seedlings of the western sand cherry *(Prunus besseyi)* can be used to produce a dwarf tree.

ARACHIS HYPOGAEA. See PEANUT

ASIMINA TRILOBA. See PAPAW

AVOCADO *(Persea)*. Propagate by T-budding or grafting the desired varieties on avocado seedlings. Both whip grafting and side grafting may be done, too. Sometimes the seedlings are raised in pots, so the grafting can be done more easily. Seeds should be planted soon after they are removed from the fruit, and not allowed to dry out.

In budding, only healthy, fat buds should be selected, and those that grow near the end of the branch are usually the best. Unlike

most other budding, the top third of the seedling should be cut off four or five weeks after the bud is inserted. The new bud should be allowed to complete a season of growth before the rest of the top above the bud is removed. Both budded and grafted trees should be staked.

Avocados are sometimes grown as houseplants by drying a fresh seed for a few days, then suspending it over a glass of water by sticking toothpicks or skewers into it. The bottom half of the seed should rest in the water with the point of the seed heading down. After roots form, it should be planted in a flower pot.

BANANA *(Musa)*. Banana trees are actually large herbaceous perennials, and the trunk is made up of compressed leaf stalks. The roots are large rhizomes which are divided to make new plants. Cut a large rhizome into pieces, each of which should weigh seven or more pounds and have at least two eyes, and plant. Sucker plants from the base of the parent may also be cut off and planted. If the leaves are shortened at planting time, water loss will be less, and the plant more likely to grow. Although a banana plant may appear to live for years, each dies after maturing its one crop of fruit, and is replaced by new sucker plants coming from the base.

BLACKBERRY *(Rubus)*. There are two types of blackberries. The upright variety spreads rapidly by sending up suckers from a vigorous root system. To start new plants, dig up these suckers and transplant them in early spring or late fall. They should be cut back nearly to the ground, after transplanting, to encourage better rooting. The trailing or vine kind of blackberry (dewberry, boysenberry, youngberry) is best propagated by tip layers.

Both types of blackberries can be increased rapidly by root cuttings,

Blackberry

although thornless kinds may revert to the thorny variety when started this way. Both can also be started from softwood tip cuttings made in early summer, which root well under mist.

BLUEBERRY, HIGHBUSH (*Vaccinium corymbosum*). Highbush blueberries are one of the most difficult small fruits to propagate. Hardwood cuttings are used by commercial nurserymen, but home gardeners usually are more successful with softwood cuttings. Rooting chemicals and a mist system are necessary.

There is a lot of argument among professionals about the best time to take the cuttings, but late spring or early summer, when growth is rapid, seems to be the best time. Use a mixture of half vermiculite and half perlite with a layer of peat moss in the bottom of the flat or pot. Some growers report better rooting by giving the plants extra hours of artificial light, so they receive about sixteen hours per day.

BLUEBERRY, LOWBUSH (*V. angustifolium*). Lowbush blueberries may be increased by softwood cuttings in the same way as the highbush berries. They can also be grown from seeds taken from the ripe

fruit. Plant these in a soil mixture of one-third sand, one-third acid soil, and one-third peat moss. Scatter the seeds on top, cover with a thin layer of peat or sphagnum moss, and keep moist. Allow the seedlings to stay in the seed bed until the following spring. If they are large enough, transplant them the following spring. If they are still weak, leave them another year, and continue the fertilizing.

Wild lowbush berries layer naturally, and both these and the seedlings may be dug up and transplanted in early spring.

BUTTERNUT (*Juglans cinerea*). Most butternuts are grown from seed. The nuts should be planted in the fall, as soon as they drop from the trees, or stratified. Although they may be planted an inch or so deep where you want them to grow permanently, it is usually better to plant them in a bed where they can be better protected from squirrels. Transplant them when they are less than three or four feet tall.

Scatter a few mothballs among the seeds to prevent pilferage until they sprout the following summer. Named varieties of butternuts may be grafted on either butternut or black walnut seedlings. A form of bark graft is usually used because of the thick bark. Because this is a tricky process for the home gardener, however, it may be better left to the experts.

CARICA. See PAPAYA

CARYA. See PECAN

CASTANEA. See CHESTNUT

CHERRY (*Prunus*). Both the sweet and sour cherries are usually propagated by T-budding the wanted variety on a seedling of either the Mazzard or Mahaleb cherry. Mahaleb rootstocks produce a more dwarf tree than the Mazzard, and in some cases they are shorter lived. Seeds may be bought from commercial seed supply houses, or the started seed-

lings purchased from tree nurseries. Seedlings for rootstocks may also be grown from seeds taken from your own cherries, or from fruit bought in stores. Be sure to plant only seeds from locally grown fruit, since those from another area may not be well adapted to your region. Cherries may also be budded on native cherries, but this is seldom satisfactory since wild trees sucker so badly.

CHESTNUT (*Castanea*). The American chestnut still survives in a few places, and seeds from these can be gathered and planted. None seem to be blight-resistant, unfortunately, so plantings are not likely to be long lived.

Most of the chestnuts being grown in this country are the Chinese (*C. mollissima*) variety, which does well wherever the climate is right for peaches. Named varieties are usually bark grafted or budded with the inverted T-bud. Regular grafting and normal T-budding do not usually work well. Seedlings of Chinese chestnuts should be used as stock, rather than hybrid seedlings.

CITRUS (*Citrus*). All of the citrus group can be grafted upon each other, or on closely related species such as kumquat. Orange, grapefruit, lemon, lime, and tangerine trees are usually T-budded on citrus seedlings, although they may also be started from softwood cuttings, leaf cuttings, regular layers, or air layers.

Seed for rootstocks should be planted as soon as possible after they have been removed from the ripe fruit, and not allowed to dry out. The seedlings can be grown in beds and transplanted into pots or rows during the spring, after frost danger is past. They should be spaced about eighteen inches apart in the rows. Budding is usually done in September and October. In Florida, the buds are commonly placed quite close to the ground, but in California they are placed higher—usually a foot above soil level. Sand is often piled over the bud the first winter to protect it.

The following spring the new bud is forced to grow by bending over the top of the seedling three inches above the bud. In late summer, after the new sprout is well started, the bent top is cut off smoothly on a slant, just above the new sprout. Leaving the top on temporarily helps feed the plant until the new tree has started. As with most other grafted trees, a stake is strongly recommended.

COCONUT *(Cocos nucifera)*. Coconuts are grown from seed. The coconut is planted only a few inches deep. Don't remove the husk. Lay it flat with the eye end slightly raised. They sprout about a month after planting, and the new shoots appear through the eye. In about a year they can be transplanted to their permanent location.

COFFEE *(Coffea arabica)*. Most coffee trees are grown from seed that is selected from trees that produce high quality beans. If the seeds are not planted at once, they should be stored in a slightly humid, cool environment, where they will keep for a few months. The seeds are usually planted under shade. They germinate in about a month, and a few weeks later are transplanted into rows in a seedling bed. After a year or two they should be planted out where they are to grow permanently.

Coffee trees can also be started by cuttings, leaf cuttings, and layers. If cuttings are used, it is necessary to use a mist system or other means of creating high humidity.

Sometimes, during periods of coffee shortages and price increases, there is interest in growing coffee in the home or in home greenhouses. Usually it doesn't work well, because the conditions that coffee trees prefer are hard to duplicate in an indoor environment.

CORYLUS. See FILBERT

CRABAPPLE *(Malus)*. Both regular and flowering crabapples are propagated primarily by budding and grafting, in the same way as apples. Some of the more vigorous varieties can be propagated fairly easily by placing softwood cuttings under mist, using a root-promoting chemical. Many propagators feel that cuttings taken from a tree started from a cutting will root better than those taken from a grafted tree.

Crabapple

CRANBERRY *(Vaccinium macrocarpon)*. The bog cranberry is a vine-like plant that roots readily from cuttings. Usually the runners or upright shoots are cut from the main plant in early summer, and stuck into a sandy soil, where they root and grow with no special care. Cuttings should be six to twelve inches long, and buried deep enough so that only an inch or so protrudes above the surface. They need a moist location, and usually root quite rapidly. By the second year the tops will begin to grow. Rooting can be done where they are to grow permanently, or they can be transplanted early in the spring.

CRANBERRY HIGHBUSH. See Viburnum, in Trees, Shrubs, and Vines section.

CURRANT *(Ribes)*. Currants start readily from hardwood cuttings. Take cuttings eight to ten inches long in late fall, and store them until spring in sand or vermiculite in a cool basement or root cellar. Then plant them outdoors and keep them moist.

Softwood cuttings also start well under mist, but most home gardeners can start all the plants needed by layering some of the outside branches of a large plant. They can also be started by stooling if more plants are wanted.

CYDONIA OBLONGA. See QUINCE

DATE *(Phoenix dactylifera)*. Either seeds or offshoots can be used to propagate dates. Grafting or budding is impossible, because the date is a monocotyledonous plant with no continuous cambial cylinder. The male and female flowers are on separate plants, and since only the females bear fruit, it is advantageous to start them from offshoots from a plant whose sex is already known. Date plantations consist mostly of female plants, since one male can pollinate many female trees. Offshoot propagation also gives one the advantage of selecting a tree of superior quality, since seedlings, even those from good trees, vary widely in their fruit-producing qualities.

It takes considerable care and skill to get the offshoot in a condition good enough to grow, and it may be best to have an experienced person guide you through this task. The soil must be dug away from the offshoot very carefully, so some soil is left attached to its roots. Then the offshoot must be cut off cleanly, not broken or pried from the parent tree, and planted before it has dried.

DEWBERRY. See BLACKBERRY

DIOSPYROS. See PERSIMMON

ELDERBERRY *(Sambucus)*. Wild elderberries can be easily started by planting seeds or dried berries when they are dead ripe, in late summer. Seedlings are also easily dug up and transplanted.

Hybrid varieties may be started from hardwood or softwood cuttings, or by digging up sucker plants. Seedlings of hybrids may not be exactly like the parent, but may still be worthwhile plants. All elderberries like a moist location.

FIG (*Ficus carica*). Figs are usually propagated by hardwood cuttings, but they may also be increased by regular layering or air layering. They can also be T-budded or patch budded on seedling trees. Seeds grow easily, but since they don't reproduce true to the parent, they are used mostly for growing understocks for budding or for developing new varieties.

Hardwood cuttings can be taken from wood up to three years old, and may be planted where they are to grow permanently. It is a common practice to place two or more cuttings in the same spot so that if one doesn't live, the others may.

FILBERT (*Corylus avellana*). Layering young trees is the best method of propagating the filbert, or hazelnut, as it is sometimes called. Young trees not only root easily, but they often have numerous branches near the ground. Filberts can be grown from seed, but the resulting trees may be quite different from the parent. They are rarely grafted or budded, and it is difficult to grow most cultivars from either hardwood or softwood cuttings.

FRAGARIA. See STRAWBERRY

GOOSEBERRY (*Ribes* or *Grossularia*). Commercially the gooseberry is propagated mostly by stools, although it can also be started from either hardwood or softwood cuttings. Most home gardeners start all the plants they want by either layering the plants or by digging up offshoots or sucker plants from around the parent bush. American varieties root readily when layered during the summer and are separated the following spring. European varieties root less vigorously, and may need an additional year to grow enough roots. Seeds are planted primarily for developing new varieties.

GOOSEBERRY, CHINESE or KIWI FRUIT (*Actinidia chinensis*). Softwood cuttings treated with a rooting compound root well under mist, and this is probably the best method of starting this plant. They may also be bud grafted or grown from root cuttings. They grow well from seed, but since the male flowers and female flowers are on separate plants, it takes years before one knows the sex of the plant.

GRAPE (*Vitis*). Most varieties root readily from hardwood cuttings. Softwood cuttings also root well under mist. Home gardeners who want only a few plants can layer some of the trailing vines in the spring. These root during the summer and may be transplanted the following spring. Certain varieties of wine grapes are traditionally grafted, because they are less productive when grown on their own roots.

For experimental purposes, seeds may be planted as soon as they are taken from the ripe fruit in the fall, and will sprout and grow well the following spring and summer.

GRAPEFRUIT. See CITRUS

GUAVA (*Psidium*). These round tropical fruits are usually started from seed. They should be taken only from the best trees, and planted in sterilized soil to reduce the chance of disease attacking the sprouting seeds. When the seedlings are about two inches high, transplant them to medium-size pots. They are ready to be planted in their permanent location when they are about a foot tall.

A few named cultivars are propagated by grafting or budding, but only with difficulty. Cuttings root poorly, although some success may be obtained by using mist and rooting chemicals. The plants can also be layered or air-layered.

HICKORY (*Carya ovata*). Although improved cultivars of hickory are sometimes grafted or budded on seedlings of wild shagbark hickory, the grafting process is difficult, and the trees started in this way tend to grow slowly and take years to bear. Most hickory trees are grown from seed.

The seed should be planted as soon as it falls from the tree, for best results. Since hickories are difficult to transplant, plant several nuts where you want the tree to grow. If more than one grows, pull out the weaker seedlings and leave only the best one. Protect the nuts from squirrels by burying a few mothballs along with them, and mulch with leaves or sawdust to prevent frost from heaving the nuts out of the ground during the winter.

JUGLANS. See WALNUT

KIWI FRUIT. See GOOSEBERRY, CHINESE

LEMON. See CITRUS

LIME. See CITRUS

LOQUAT (*Eriobotrya japonica*). This warm-weather plant from China grows easily from seed, which should be planted directly from the fresh fruit. Trees used only for ornamental purposes are grown from seed, as are seedlings used as understocks. Improved varieties of loquats, however, are T-budded or whip-grafted on seedlings, or grafted on an older tree. Loquats can also be grafted on quince seedlings, resulting in a dwarf tree. They may also be propagated by air layering, using rooting chemicals on the layered part.

MALUS. See APPLE

MANGO *(Mangifera indica).* This tropical fruit is usually grown from seed, but some named cultivars are propagated asexually.

Plant the seeds in a sterile medium as soon as you remove them from the fruit. If fast sprouting is desired, remove the tough outer endocarp before planting, and the seed should sprout in two or three weeks. When it is well established, transplant it to a pot or other container.

To propagate an improved variety, veneer-graft or chip-bud it to a seedling that is still young and growing vigorously. If you do budding, a week or so after inserting the bud, cut off the top of the seedling a few inches above the bud to force it to "break." After the growth from the bud is about four inches long, the rest of the wild seedling is cut off just above the new bud.

Mangos can be air-layered, and may also be propagated by cuttings, but only with difficulty.

MULBERRY *(Morus).* Propagate the native varieties of mulberries by planting their seeds. To propagate named varieties, take either softwood cuttings or hardwood cuttings from wood grown the previous season. Weeping mulberries are grafted on mulberry seedlings. Usually seedlings grown from the white mulberry *(M. alba),* a native of China, or the red *(M. rubra),* an American native, are considered the best for grafting.

MUSA. See BANANA

NECTARINE. See PEACH

OLIVE *(Olea europaea).* Olives are usually grown from hardwood or semihardwood cuttings. Rooting is much improved if mist and rooting chemicals are used.

Propagation of good trees may also be done by grafting scions on small olive trees grown from seed. The seed often takes a year or more to sprout, though chipping the hard shell helps somewhat.

Germination is erratic, so plant extra seed, since not all of it will grow.

Sometimes in the spring large chips are cut from olive trees and buried, bark side up, in the ground. If kept moist and warm, they begin to sprout and grow, although the practice is hard on the parent tree.

ORANGE. See CITRUS

PAPAW *(Asimina triloba).* These are best started from seed which is taken from the fruit as soon as it is ripe, and planted at once. Otherwise, stratify for three to four months at 40° F. and plant in the spring. They are best grown in partial shade, and take a long time to start. They should be planted in their permanent location at an early age, because the long taproot they develop makes them hard to move later on.

PAPAYA *(Carica papaya).* Outstanding varieties of papaya are grafted on papaya seedlings. They may be propagated by cuttings, too, if bottom heat is used, but the most common way of starting papayas is by seed. The seed can either be taken from the ripe fruit, or stored dry for a time before planting. Home gardeners may want to start seedlings in pots to save the trouble of transplanting. Plant several seeds in each pot of sterilized soil or artificial soil mix. It is important that it be sterile, because papaya seeds are very susceptible to soil disease when they first sprout. When the seedlings are a few inches tall, save the strongest one and pinch off the rest. In a short time the young trees may then be planted directly to the field with no disturbance to their root system.

PASSION FRUIT *(Passiflora edulis).* This fruit is sometimes propagated by cuttings, layering, root cuttings, and grafts, but mostly it is grown from seeds. The seeds germinate easily, often two or three weeks after planting.

PEACH and NECTARINE *(Prunus persica).* Although their seeds have a better record of producing worthwhile fruit trees than the seeds of most other fruits, peach and nectarine trees grown from seedlings do not reliably produce fruit similar to that of the parent. They are usually grown by budding good varieties on peach, apricot, or plum seedlings.

Peach

Peach seeds are usually stratified for three or four months before planting, but it is also possible to grow seedlings by planting the seeds immediately after separating them from the ripe fruit. Peach understocks grow best in light, sandy soils that warm up fast, and plum understocks grow better in heavier soils. The western sand cherry *(P. besseyi)* is sometimes used as an understock for certain cultivars to produce a dwarf tree. Some gardeners have reported success in grafting peach on the Nanking cherry, for a tree that is even more dwarf. Although a peach tree can also be grafted on an almond seedling, resulting in a dwarf tree, it is likely to be short-lived.

The T-bud method is usually used for bud grafting. Budding is done in late August in northern areas and in June in the South.

It is difficult to start peach cultivars from cuttings ordinarily, but some varieties root fairly well from softwood cuttings that have been treated with rooting chemicals and placed under a mist system. Even hardwood cuttings are sometimes successful where the growing season is extra long.

Many horticulturists recommend that you do not plant a new peach tree near the spot where an old one was taken out, since many diseases that affect the roots of peaches are inclined to live on in the soil. If such planting must be done, it is better to graft the new trees on plum roots.

PEANUTS (*Arachia hypogaea*). The peanut is neither a nut or fruit, but a member of the pea family. It is best grown where seasons are rather long, because it is very susceptible to frost. Peanuts are always grown from seed (the unroasted nuts, of course). The female flowers, after fertilization, develop pods which bury into the soil. There they complete their development. They like a sandy soil that is not acid.

PEAR (*Pyrus communis*). Some pear varieties, such as Bartlett, can be started by hardwood cuttings, or softwood cuttings under mist, and with the use of rooting chemicals, but only with difficulty. The most common method of propagation, and easiest for the home gardener, is by T-budding or cleft grafting.

Commercial growers use a variety of seeds for growing seedlings for grafting stock, some of which are selected because they are resistant to fireblight. Home gardeners plant seeds from pears that grow locally, or buy seeds that are grown commercially. Calleryana is a blight-resistant variety, but the seedlings produced are less hardy than those grown from Bartlett seeds.

Pears may also be grown on quince roots, which produce dwarf trees. Because some cultivars don't graft well upon quince seedlings, they are first grafted to a variety such as Old Home, allowed to grow

a season, and then regrafted with the variety wanted. Pears may also be successfully grafted on apples, but the tree usually dies within a few years.

PECAN (*Carya Illinoensis*). Pecan cultivars are usually grafted or budded on seedling trees. Patch budding and ring budding are the most common methods for pecan propagation.

Pecan seedlings grown from seeds gathered from healthy, vigorous-growing wild trees make the best rootstocks. They should be planted immediately after harvesting, or held at near-freezing temperatures for four months and planted in the spring. Plant the seeds in sandy soil and keep them moist and shaded. The second year they may be budded with the desired variety. The patch bud is as effective as the ring method, and considerably easier.

Pecans are sometimes grafted on hickory seedlings, too, but the nuts do not grow to normal pecan size.

PERSEA. See AVOCADO

PERSIMMON (*Diospyros*). Both the native (*D. virginiana*) and the Oriental (*D. kaki*) are usually propagated by grafting or bud grafting the named cultivars on wild seedlings of *D. virginiana* or *D. kaki*. Cleft grafting is usually the most successful on Oriental varieties. The seeds must be picked as soon as they are ripe, and stratified for over the winter at 40 degrees F., unless they are planted as soon as the fruit ripens. If the seed has dried out, it should be soaked for a day or two before being stratified. Even under the best conditions, germination is likely to be slow.

Newly grafted trees can be moved to their permanent location after the graft has grown for a season. Larger trees should have their taproot cut off a foot or so below the surface of the soil a year before moving, to encourage side root growth.

PINEAPPLE (*Ananas comosus*). Pineapples are herbaceous monocotyledons and may be started by crowns, by slips that develop on the stalk just below the fruit, and by root suckers that grow below ground level or in leaf axils. The root suckers are harder to dig off and plant, but they produce fruit much faster than either slips or crowns. They are planted about twenty-two inches apart and six inches deep.

To grow pineapples as houseplants or in a greenhouse, cut the crown off a pineapple fruit, plant it in a pot of well-drained soil, and fertilize it well. A crown takes nearly two years to grow and mature a ripe fruit.

Pineapples respond to fertilizer, particularly liquid feeding, and for best results should be fed regularly. They are extremely sensitive to frost.

PISTACHIO. See PISTACIA in Trees, Shrubs, and Vines section.

PLUMS (*Prunus*). The most common and satisfactory way of propagating plums is by T-budding. Named cultivars are budded on seedling plums or clones of plums grown especially for rootstocks. Although in theory all plum varieties can be grafted or budded on each other and on peaches, almonds, cherries, and apricots, in practice, certain plums grafted on other plums do not succeed. Within the

Plum

plum family there are three groups, and, as a rule, it is advisable to bud the Japanese varieties on peach seedlings or seedlings of Japanese plums, European on *P. cerasiera* or on European plum seedlings, and American hybrids on wild native seedlings. Plums may also be grafted on the western sand cherry *(Prunus besseyi)* to produce a dwarf tree, but this should be considered only for a garden tree, since it may be short-lived.

Like cherries, many plums are infected with various kinds of virus that shorten their life, especially in the colder regions.

POMEGRANATE *(Punica granatum).* The pomegranate can be propagated easily by either hardwood or softwood cuttings. The use of a rooting compound helps greatly, and the softwood cuttings must be misted or kept in a high humidity. Root suckers can be dug from the base of the plant and transplanted, and it is also possible to increase them by layering, budding, or grafting. They also grow easily from seed, but the resulting shrub or tree is always always of inferior quality.

PRUNUS. See PLUM, PEACH, ALMOND, CHERRY, APRICOT

PYRUS COMMUNIS. See PEAR

QUINCE *(Cydonia oblonga).* Home gardeners find that layering is the easiest way to propagate the quince. Hardwood cuttings, with a bit of two-year wood attached to the base of the cutting, root well and grow rapidly. Quince rootstocks for pear trees are sometimes produced in this way. Quince may also be budded or grafted on quince seedlings or rooted cuttings of wild quince.

For flowering quince, see the section on Trees and Shrubs.

RASPBERRY *(Rubus).* Red raspberries are propagated from the suckers that grow vigorously both inside and outside the rows. Either one- or two-year-old plants can be cut back nearly to ground level and transplanted in spring or fall. New sprouts can be transplanted during a rainy week in early summer if kept watered. A plant will produce even more suckers if you cut completely around it, with a spade, about a foot from the stems.

Root cuttings can also be used for propagating. Dig the entire plant in early spring, cut the roots into pieces two inches long, and plant horizontally one-half inch deep in sandy soil. Transplant the following spring.

Black raspberries are rooted by tip layering. Although the plants do this naturally, if more plants are wanted, you can help out nature. Bend over some of the long new canes in mid-summer, and cover the tips with five or six inches of soil. Roots will form in the soil and new shoots will grow from the tips. The new plants may be severed and transplanted the following spring. Some gardeners sink a pot of soil in the ground where they make the tip, so the new plant will be formed in a pot. These can be dug and planted in a new location the same year, thus gaining a season.

Black raspberries also can be propagated by softwood cuttings taken from the new shoots that come from the ground in early summer, which root easily under mist.

RIBES. See CURRANT, GOOSEBERRY

SAMBUCUS. See ELDERBERRY

SASKATOON *(Amelanchier alnifolia).* Seeds may be used to propagate wild varieties, and even named varieties produce a large percentage of good plants from seed. Plant the seed as soon as it is ripe, or stratify it over the winter at 40° F. and plant it in the spring. Cuttings are possible, but not easily, and the plants layer well. Some of the best varieties are bud grafted on wild seedlings.

STRAWBERRY *(Fragaria).* Because the strawberry forms runners so readily, digging them is the easiest way for home gardeners to start additional plants. Most commercial plants are started by this method, too, although tissue culture is becoming more popular. New strawberry beds are usually started in the spring, and the plants are dug at that time. If you want to gain a season, fill three-inch plastic pots with soil and set in your strawberry patch. Steer the new runners to them and let them form roots in the pots. In late summer, cut the runners apart, move the pots to a convenient place where you can keep them watered and fertilized for two or three weeks, and then plant them out. They will bear fruit the following year.

A few strawberry plants of exceptional quality have gone off the market because they produced so few runners that nurseries found them unprofitable. If you are lucky enough to have any of them, you can start more plants by carefully digging the whole plant and separating the crown. Each piece of crown will grow into a new plant.

TANGERINE. See CITRUS

VACCINIUM. See CRANBERRY, BLUEBERRY

VITIS. See GRAPE

WALNUT *(Juglans).* Persian, Carpathian, or English walnuts are different names for *J. regia,* and are easily grown from seed. Improved varieties are whip grafted on one-year seedlings of various strains of *J. regia* or black walnut *(J. nigra).* Bark or patch grafting is the usual method of top-working larger trees.

Black walnuts are propagated the same as butternuts.

TREES, SHRUBS, AND VINES

The following list includes ornamental and forest trees, flowering shrubs, evergreens, and woody vines. Fruits, perennials, and herbaceous plants are listed elsewhere. Because the number of trees, shrubs, and vines is nearly unending, this list includes only those that are most widely grown and propagated. Stratification times given are minimum, and longer periods may be used.

ABELIA *(Abelia).* Usually propagated by hardwood cuttings. Softwood cuttings treated with a rooting compound and placed under mist or plastic will also root. These should be taken in late summer when the new growth is slightly hardened. Seeds may also be planted when they are ripe.

ABIES. See FIR

ABUTILON *(Abutilon)* FLOWERING MAPLE. This plant is often grown as a houseplant where it cannot be grown outdoors. It can be started by softwood cuttings placed under plastic or mist. A rooting compound helps. It may also be grown from seeds.

ACACIA *(Acacia).* Most varieties of acacia will grow from softwood cuttings of partially matured wood, or from root cuttings, but it is usually started from seeds. Because the seed coats are very hard, the usual practice is to soak

the seeds in concentrated sulfuric acid for a half-hour or so, and then wash them thoroughly directly after soaking. They can also be softened by pouring boiling water over the seeds and leaving them in the water for the next ten hours or so as it cools. Acacias should be transplanted while the plants are still small because they form a taproot at an early age which makes them difficult to move.

ACER. See MAPLE

AESCULUS. See BUCKEYE

AILANTHUS *(Ailanthus altissima)-* TREE OF HEAVEN. The sexes of this species are on separate trees, and since the male trees smell bad, planting the females is a real advantage. To propagate female trees, take root cuttings in the spring, or dig up suckers from around them. Scions from female plants may also be grafted on seedlings.

Although propagation by seed is simple, and the trees often self-sow, the seeds will be approximately half male. Seeds should be stratified as soon as ripe, 40° F. for two months, or planted as soon as ripe. Scions from female plants may also be grafted on seedlings.

ALBIZZIA *(Albizia julibrissin)* SILK TREE. Root cuttings taken in the spring are usually quite successful. Seeds are a more common method

of propagation, however. Because they have a hard seed coat, the seeds should be treated in the same way as those of acacia.

ALDER *(Alnus).* The many types of alder may be started from layers, suckers, or even from hardwood cuttings. Seed is the most common way. The European alder *(A. glutinosa)* can be planted in the fall as soon as it is ripe, but other varieties should be stratified at 40 degrees F. for three months and planted in the spring.

ALLAMANDA *(Allamanda).* This tropical vine is sometimes grown in large tubs in the north. It can be propagated by both hardwood and softwood cuttings, and rooting compounds are advised. Taking a bit of old wood along with the new growth on a softwood cutting often helps rooting.

AMELANCHIER. See SERVICEBERRY

ANDROMEDA. See PIERIS

ARALIA *(Aralia).* These shrubs or trees are best propagated from seeds planted in the fall as soon as they are ripe. Soak in sulfuric acid for one-half to one hour, wash in water, and plant at once. Aralia may also be propagated by root cuttings or by root graftings or budding.

ARAUCARIA HETEROPHYLLA. See NORFOLK ISLAND PINE

ARBORVITAE *(Thuja)*. The American arborvitae (*T. occidentalis*), native white cedar is easily propagated from seed sown in either fall or spring. Seed kept until spring should be stratified for two months at 40° F. for best results.

Named varieties of arborvitae — the globes, pyramids, and other kinds—may be propagated by hardwood cuttings taken in midwinter and rooted under mist with bottom heat. Softwood cuttings taken early to mid-summer will also root under mist. In both cases, the use of rooting chemicals is beneficial. Fairly long cuttings of even a foot or more in length will often root in this manner.

Spreading varieties or those with branches near the ground can be layered if only a few plants are wanted. It may take a year or more for a good root system to develop, however.

Oriental arborvitae (*T. orientalis*) roots less readily from cuttings than the American variety. Softwood cuttings seven inches long, taken in early summer, which are treated with a rooting chemical and placed under mist, root fairly well. Side grafts of Oriental cultivars are sometimes made in the greenhouse in winter on potted seedlings of *T. orientalis*, and the plants are transplanted in the spring.

ARBUTUS. See MADRONE, PACIFIC

ASH *(Fraxinus)*. Sow the seeds as soon as they fall from the tree. Older seeds germinate poorly. Two or more months of stratification at 40 degrees F. is beneficial, if planting cannot be done at once. Some varieties may take two years to germinate. Named varieties, such as the seedless kinds, are often grafted on ash seedlings.

ASPEN. See POPLAR

AUTUMN OLIVE. See ELAEAGNUS

AZALEA *(Rhododendron)*. Deciduous azaleas are commonly started from seeds because that is the easiest and fastest method. Seeds should be gathered as soon as they ripen and planted immediately, or stored in a cool place in airtight containers until spring. They may be planted either in outdoor beds or in flats in the greenhouse. If planted in flats in the fall, they should be kept cold until mid-winter or late spring, then brought into a warm area and allowed to germinate. In the bright spring days they will grow rapidly and will be ready to bloom more quickly than those grown in outdoor beds.

Azalea

Deciduous azaleas can also be started easily by simple layering, trench layering, or air layering, but the rooting process takes longer than with most flowering shrubs.

Softwood cuttings do not root easily, and the leaves are likely to drop off the plant before it has rooted. Some gardeners have been successful in propagating them under plastic, by keeping the medium very wet and misting them once or twice a day. Use a mixture of half vermiculite and half perlite over a layer of peat moss in three-inch pots. Dip the cutting in a rooting chemical and insert it so it just touches the peat moss. Bottom heat is often beneficial.

Named varieties of azaleas are also grafted on azalea seedlings. Pot the seedlings in the spring and do the grafting during the late summer in a greenhouse. A side graft, similar to that used in grafting conifers, is the common practice.

Evergreen azaleas are also propagated by seeds, layers, grafting, or suckers. The seed pods are picked in late fall, then dried. The seeds are usually planted in the greenhouse during the winter or spring in a flat half filled with peat or shredded sphagnum moss, and covered with a thin layer of vermiculite. They germinate best at a temperature of approximately 70 degrees F. Avoid watering them with hard (alkaline) water, because any excess of lime will damage the seedlings. Rain water or water from a dehumidifier may be used instead, if local water is unsuitable.

Evergreen azaleas may also be started from cuttings. The use of rooting chemicals and mist is beneficial, but overmisting should be avoided. The cuttings may be taken any time after growth starts, but they often root best if taken in late summer, after the wood has begun to harden, and placed in a greenhouse.

Be careful not to use lime or hardwood ashes on any young azalea plants. Both the evergreen and deciduous azaleas are acid-loving plants.

BALD CYPRESS. See TAXODIUM

BAMBOO *(Arundinaria, bambusa)*. Of the hundreds of varieties of this tropical plant, only two are native to the United States. Bamboo is propagated by dividing the clumps just before the period of fastest growth begins.

BARBERRY *(Berberis)*. Barberry can be easily propagated either by dividing the clumps or layering some of the lower branches. They can be started in greater numbers by softwood cuttings taken in late spring or early summer when growth is the fastest. These should be treated with rooting chemicals and put under mist.

Common varieties of barberry are easily grown from seed that has been collected or purchased. Fall is the best time for planting, but if the seeds are stratified for four to eight weeks at 40° F., spring planting will give good results, too.

BASSWOOD. See LINDEN

BAYBERRY. See MYRICA

BEECH *(Fagus)*. Improved cultivars are grafted, usually by the cleft-graft method, on seedlings of the same species. If the beeches you intend to propagate are not named varieties, however, the easiest way to start them is from seeds planted in the fall. Seeds saved for spring should be stratified for three months at 40 degrees F., and not allowed to dry out. The trees should be planted in their permanent location at an early age, before they develop their long taproot which makes digging and moving so difficult.

BERBERIS. See BARBERRY

BIRCH *(Betula)*. Weeping varieties are propagated by grafting on European birch seedlings *(B. verrucosa)*, and other varieties are grafted on seedlings of *B. pendula* or *B. papyrifera*. Paper white birch and other common species are started from seed collected and planted in the fall or late winter. Seeds may be planted outdoors in beds or in flats and allowed to freeze for several weeks before germinating, stratified for two months at 40 degrees F. or stored in airtight containers in a cold place and planted in the spring.

Softwood cuttings will root fairly well if they are treated with root chemical treatment and placed under mist. Low-growing limbs on young trees can be layered successfully, also.

BITTERSWEET *(Celastrus)*. Like most vines, bittersweet is easily started from layers or compound layers. Hardwood, softwood, or root cut-

tings can also be used. The softwood ones root readily under mist.

Since the sexes are on separate plants, and only the females produce berries, asexual methods are preferred over propagation by seeds. For mass plantings, seed is often satisfactory, however, and fall is the best time for planting. If seed must be held over, it should be stratified for three months at 40 degrees F.

BOSTON IVY. See PARTHENOCISSUS

BOTTLEBRUSH *(Callistemon)*. Softwood cuttings root easily and are the preferred method of rooting the most attractive varieties. Seeds also grow easily, but the resulting seedlings vary a great deal, and many poor specimens are likely to develop from them.

BOUGAINVILLEA *(Bougainvillea)*. Hardwood cuttings set in sand, or softwood cuttings treated with rooting compounds and placed under mist are the most common ways of starting this colorful vine. Layers and air layers can also be used.

BOXWOOD *(Buxus)*. Hardwood cuttings taken in the fall are the most common method of propagating boxwood. These can be rooted in a greenhouse or cold frame over the winter. Softwood cuttings will root under mist. Those cuttings eight inches or more in length seem to root best and make the huskiest plants. Boxwood can be grown from seed as well, but the seedlings grow so slowly that the method is seldom used. If seeds are used, they should be sown as soon as ripe, or else stratified.

BROOM *(Cytisus)*. Softwood cuttings treated with rooting chemicals root easily, and for propagating the best plants, this is the most reliable method. Hardwood cuttings also root well, and named varieties may also be grafted using Spike Broom or Scotch Broom seed-

lings as stocks. Seeds may not always come true, especially if more than one kind of broom is growing in an area. If seeds are used, they will germinate faster if the hard seed coats are treated with sulfuric acid. Soak them for one-half hour, then rinse thoroughly with water just before planting.

BUCKEYE *(Aesculus)*. HORSE CHESTNUT. The easiest way to start this tree is to plant the seeds soon after they fall, or stratify them for four months and plant them in the spring. Low-growing branches on young trees can be layered. Cultivars are sometimes propagated by grafting them on seedling roots using either the whip graft or T-bud.

The dwarf buckeye *(A. parviflora)* can be grown from root cuttings, or seeds sown in summer as soon as they are ripe.

BUCKTHORN *(Rhamnus)*. Some varieties self-sow easily, and sometimes seedlings that have sprouted from bird-dropped seeds can become real weeds. Collecting these seedlings is often an easy way to get new plants. Seeds can be planted in the fall or stratified at 40 degrees for three months and planted in the spring. Selected cultivars can be propagated under mist from either hardwood or softwood cuttings.

BUTTERFLY BUSH *(Buddleia)*. Softwood cuttings root well if taken in early summer and placed under plastic or mist. Hardwood cuttings are also often used. Seeds can also be used to propagate buddleia, but seedlings vary in quality. Young plants—both seedlings and cuttings—may need to be protected over the winter where cold temperatures are likely.

BUXUS. See BOXWOOD

CALLISTEMON. See BOTTLEBRUSH

CALLUNA VULGARIS. See HEATHER

CAMELLIA *(Camellia)*. Layering, seeds, grafting, and cuttings are all used for starting new camellias. Asexual means must be used for propagating named varieties, and seeds are used mostly for developing new kinds. For best results, the seeds should be planted as soon as they are ripe, before the seedcoat hardens and becomes dry. Even under ideal conditions, camellia seedlings take many years to bloom.

If only a few plants are wanted and there are branches near the ground, they can be layered. Young, small branches root best, but even these may take two years to form a good root system. Air layers are practical, too, to propagate only a few plants.

Most varieties of camellia can be propagated by cuttings. The common camellia *(C. japonica),* roots quite easily when taken in mid-summer, when the new growth is slightly matured. Rooting chemicals are advisable, and the cuttings should be placed under mist.

Camellias are also propagated by grafting named varieties on small seedlings. The side or veneer graft is most commonly used, and sometimes soil is piled high around the seedling, covering the graft union, to protect it. A shaded glass jar is used to cover the plant, and it is removed after the scion has started to grow.

CAMPSIS. See TRUMPET-CREEPER

CATALPA *(Catalpa)*. Seeds start easily and grow rapidly, whether planted in flats or in the ground. They may be planted either in fall as soon as they are ripe, or in spring after having been stored in a tight container in a cool, dry place.

The cultivar *C. bignonioides,* Nana, is sometimes grafted on a seedling of *C. speciosa* to create the umbrella tree. The seedling is allowed to grow to a height of about six feet, then bud grafted with several

buds, or grafted in spring. Hardwood cuttings are sometimes used for propagating, as are softwood cuttings under mist.

CEDAR *(Cedrus)*. Cuttings taken in late summer can be rooted in a greenhouse over the winter with the help of bottom heat, but they do not start easily. A rooting chemical is advised. Seeds planted as soon as they are harvested grow easily, or they can be stratified for two months or more at 40 degrees F. Improved cultivars of cedar may be grafted on seedlings. Cedrus is the only true cedar and grows well only in mild climates. The term cedar is commonly used throughout the country to describe other evergreens, including juniper, arborvitae, and cypress.

CELASTRUS. See BITTERSWEET

CELTIS. See HACKBERRY

CERCIS. See REDBUD

CHAENOMELES. See QUINCE, FLOWERING

CHAMAECYPARIS *(Chamaecyparis)*. FALSE CYPRESS. Cuttings taken in late fall can be rooted in a greenhouse with bottom heat and rooting chemicals. Seeds should be dried after harvesting, and stratification for sixty to ninety days will speed up germination.

Improved varieties may be grafted on *Thuja occidentalis* or *Cyprus lawsoniana* seedlings.

CHERRY, FLOWERING (Also FLOWERING ALMONDS, FLOWERING PLUMS and **REDLEAVED CHERRIES** and **PLUMS)** *(Prunus)*. Softwood cuttings of most varieties root well under mist if rooting chemicals are used. Some varieties reproduce true from seed if there are no other kinds of cherries around to cross-pollinate them.

T-budding is the usual method of propagating named varieties, however. Seedlings of the Mazzard cherry are suitable as understock. Wild cherry and plum seed-

lings or suckers can also be used, but these sucker very badly, and are seldom satisfactory.

CHINABERRY. See MELIA

CHIONANTHUS. See FRINGETREE

CLEMATIS *(Clematis)*. Clematis vines can be started from seeds, but the seeds should be stratified for two or more months before planting. Because the large flowering varieties do not come true from seeds, they are usually propagated from softwood cuttings taken in spring. These root well under mist if rooting compounds are used. Cuttings of partially matured new growth, taken after blooming, can also be used. These should be treated the same way as softwood cuttings. They may need additional light and heat to develop a good root system before winter, and if root growth has been light, may also require the protection of winter storage in a greenhouse or cold frame the first year.

Clematis may be started from leaf cuttings, which root easily. A small number of plants can be grown from each plant by layering, as well. Choose wood that has wintered over, and layer it early in the spring. Cover the layered area with a mulch of leaves or grass clippings to keep it cool and moist. Rooted plants should be dug up and separated the following spring.

CORNUS. See DOGWOOD

COTINUS COGGYGRIA. See SMOKETREE

COTONEASTER *(Cotoneaster)*. Although these are difficult to start from seed, germination can be hastened by soaking the seeds for an hour or so in sulfuric acid, washing them thoroughly, and then stratifying them over the winter at 40 degrees F. before planting in the spring. If desired, instead of the acid treatment, they may be "after ripened" by stratifying them at a temperature of 75 degrees F. for four months, followed by normal stratification.

Softwood cuttings of most varieties work well. Usually they do best if they are taken a bit later in the season than most cuttings, after the wood has hardened somewhat, and placed under mist. Rooting chemicals help also. Layering is a good way to start a small number of plants.

COTTONWOOD. See POPLAR

CRABAPPLE, FLOWERING *(Malus).* The method of starting regular crabapples and flowering crabs is the same as described for apples in the fruit section. They can be grafted by most of the methods mentioned, but the cleft graft, whip graft, and T-bud are most popular. Seedlings grown from either regular apple or crabapple seed can be used as understock, or they can be grafted on cloned dwarf rootstocks, such as the Malling series.

A few cultivars of flowering crabs can be started by softwood cuttings taken in early summer. Treat them with rooting chemicals and put them under mist. It is important to remove the cuttings from the mist as soon as they begin to root, or they will deteriorate quickly. It will probably be necessary to keep the newly rooted plants in a warm greenhouse for a few additional months and overwinter them in a cool greenhouse or protected cold frame, to develop good root systems.

The callused air layer method (see Cuttings) is effective for rooting small numbers of trees.

On young trees, branches that are near the ground may be layered, but it usually takes at least two years for a good root system to develop.

CRAPE MYRTLE *(Lagerstroemia indica).* Leaf cuttings under mist is one of the best methods of propagation. They should be started in small pots rather than in flats or

the ground because they are difficult to transplant. They can also be propagated by seeds and softwood cuttings.

CRATAEGUS. See HAWTHORN

CRYPTOMERIA, JAPANESE *(Cryptomeria japonica).* Cuttings of partially mature softwood can be rooted with the help of bottom heat and a rooting compound. They need extra light as roots begin to form, to stimulate heavier root growth. They can also be started from hardwood cuttings and by seeds, which should be planted as soon as they are ripe and before they dry out.

CYPRESS *(Cupressus).* Stratify seeds for a month before planting. Cuttings can be rooted if taken in the winter and treated with rooting chemicals and bottom heat in a greenhouse. Named cultivars can be grafted on cypress seedlings. This is also best done in a greenhouse on potted seedlings with a side or veneer graft.

CYTISUS. See BROOM

DAPHNE *(Daphne).* Softwood cuttings root well when, after being treated with rooting chemicals, they are placed under mist. Some varieties root better if the cuttings are taken later in the season after the wood is partially mature. Layering some of the side branches in very early spring is also useful, although it may be necessary to wait a year or more to ensure enough roots to transplant the new plant safely. Both the lilac daphne and the February daphne can be started from root cuttings. They can be grown from seeds, but with difficulty.

DEUTZIA *(Deutzia).* Softwood cuttings root easily under mist or plastic. Hardwood cuttings taken in fall or winter also root well and can be planted directly in the ground in spring. Deutzias can also be divided.

DOGWOOD *(Cornus).* There are dozens of different kinds of dogwood throughout the country, and most of the native varieties can be started easily from seeds or division. The popular flowering dogwood (*C. florida*) can be started by layering and division but is usually propagated by softwood cuttings taken just after the flowering season is over. Treatment with rooting chemicals helps, as does the use of mist. The hardy Siberian dogwood (*C. alba*) starts well from hardwood cuttings taken in early spring. Some varieties, such as the red flowering (*C. florida rubra*), are sometimes propagated by T-budding in the summer. Commercially they are often whip grafted in the greenhouse during the winter.

DOUGLAS FIR *(Pseudotsuga menziesii).* Seeds should be sown in the fall soon after ripening, or stratified for two months at 40 degrees F. Cuttings taken from young trees in late winter can be rooted but they do not form roots easily. Rooting compounds and bottom heat help; and the cuttings taken from trees that were, themselves, started from cuttings, usually root easier. Cultivars are sometimes grafted on wild seedlings.

ELAEAGNUS *(Elaeagnus).* Russian olive (*E. augustifolia*) can be started by division, layers, root cuttings, and hardwood cuttings, but seeds are the method of propagation most frequently used. The seeds are usually soaked in sulfuric acid for a half-hour, then stratified for three months and sown in early spring. Hardwood cuttings taken in early spring root well, especially when treated with rooting chemicals. Softwood cuttings are more difficult to root, but possible, under mist.

Autumn olive (*E. umbellatus*) is hardier than the Russian, and is propagated in a similar fashion.

ELDER (*Sambucus*). The elder grows well from seeds that should be stratified for three or four months, but if named varieties are wanted, they should be propagated asexually. Softwood cuttings taken in early summer root so easily that this method is most common. For a limited number of plants, however, suckers may be dug from around the old plant and transplanted in early spring.

ELM (*Ulmus*). Seed planted as soon as harvested, or stratified seed sown in the spring, is the usual form of propagation. Seed does not keep long in ordinary storage, but can be kept for several years if in sealed containers at near freezing temperatures. Some of the outstanding varieties of elm are grafted on seedlings of the same species.

ENGLISH IVY (*Hedera helix*). Cuttings root easily from this vine, or it can be layered.

ERANTHEMUM (*Eranthemum*). These warm-climate plants may be started by cuttings taken from young new growth.

ERICA. See HEATH

EUCALYPTUS (*Eucalyptus*). Seeds are the usual form of propagation. Sow in sandy, sterilized soil, and transplant the seedlings to pots, since they are difficult to transplant when larger. Although it is possible to propagate eucalyptus by cuttings, they root with difficulty even when given special care.

EUONYMUS (*Euonymus*). Seeds may be planted as soon as they are ripe or after being stratified for three or four months at 40° F. Named cultivars must be started asexually, however. Hardwood cuttings, taken in early spring, are an easy way to propagate the deciduous types, and softwood cuttings are successful, too, using rooting chemicals and mist. Evergreen varieties are best rooted from softwood cuttings taken after the wood has begun to harden. Layering is possible for all species.

EUPHORBIA. See POINSETTIA

FAGUS. See BEECH

FALSE CYPRESS. See CHAMAECYPARIS

FIR (*Abies*). Planting seeds in fall, as soon as they are ripe, is best because fir seeds lose their viability fairly quickly. (See Evergreens in Seed chapter.) Fir seedlings are particularly susceptible to the damping off diseases soon after they germinate, so plant in sterile soil or mix, and use a good fungicide. Named varieties of firs, including the dwarf varieties, can be rooted from softwood cuttings, but only with difficulty. Coarse gravel has been used as a rooting medium with good results. Bottom heat in a winter greenhouse is almost a necessity when rooting hardwood cuttings, as are rooting chemicals. Most nurserymen feel that cuttings taken from fairly young plants that were themselves started from cuttings, root easiest. Air callused cuttings are worth trying, also.

Named varieties are sometimes grafted, and if so, terminal tips should be used as scions, because they are more easily trained into a tree shape.

FIRETHORN. See PYRACANTHA

FORSYTHIA (*Forsythia*). Most varieties can be started by hardwood cuttings taken in winter, stored in sand and vermiculite for a few weeks, and planted out in early spring. All kinds seem to start well with softwood cuttings taken in late spring and early summer and put under mist. Rooting chemicals help with both methods.

Larger shrubs may be divided, and offshoots may be dug up from around them.

FRAXINUS. See ASH

FRINGE TREE (*Chionanthus virginicus*). Fall-planted seed works well, but often doesn't germinate until the second spring. Stratification of the seed for a month at 60 degrees F. followed by two months of stratification at 40 degrees F. speeds up germination. Softwood cuttings root with difficulty but are possible. Take them in late spring, use rooting chemicals and mist and a mixture of perlite and vermiculite as a medium. Sometimes they are grafted on white ash seedlings.

GARDENIA (*Gardenia jasminoides*). Hardwood cuttings taken during the winter can be rooted in a greenhouse using rooting chemicals and bottom heat. They should be shaded from direct sun, however, and kept under glass, plastic, or mist until they root. A mixture of half perlite and half peat moss is a good rooting medium.

GINKGO (*Ginkgo biloba*). Seeds are a common way of starting this tree, and the seeds should be harvested as soon as ripe and not allowed to dry out. They should be cleaned and stratified at 40° F. for about three months before planting. Although seeds are the usual method of propagation, cuttings are preferable because the plants are either male or female, and the female gives off an unpleasant odor. To propagate the males asexually, take softwood cuttings in mid-summer. They root quite well, but extra care with watering and fertilizing must be taken to get them growing well enough to make a good plant before winter. Named varieties can be grafted or budded on seedlings.

GLEDITSIA TRIACANTHOS. See HONEY LOCUST

GOLDEN-RAIN TREE (*Koelreuteria*). The seeds germinate faster if they are soaked in sulfuric acid for sixty minutes, rinsed thoroughly, and stratified for three months at 40 degrees F. Softwood cuttings taken in early spring root well if treated with rooting chemicals and put under mist. Layers and root cuttings are useful methods of propagation if only a few plants are wanted. Outstanding varieties are grafted or budded on seedlings.

HACKBERRY (*Celtis*). Seeds may be planted in the fall, as soon as they are ripe, or stratified over the winter and planted in the spring. Outstanding trees are sometimes propagated by grafting or chip budding on seedlings of the common hackberry. Cuttings are possible, but difficult.

HAMAMELIS. See WITCH HAZEL

HAWTHORN (*Crataegus*). Cultivar varieties are T-budded or grafted on hawthorn seedlings, since they will not come true from seeds.

Plant the seeds in mid-summer directly after ripening, and they will germinate the following spring. If they are not planted right away, seeds should be stratified over the winter at room temperature, then stratified for another four or five months at 40 degrees F. Then plant the seeds in mid-summer. An alternate method is to soak the seeds in sulfuric acid for an hour or so, and stratify for 3 months at 40 degrees F. Some varieties root well from root cuttings taken in early spring.

HEATH (*Erica*) and **HEATHER** (*Calluna vulgaris*). Cuttings of these two similar plants root easily under mist in early summer. Rooting chemicals help. Mature plants may be divided if only a few new plants are needed.

Although the best varieties must be propagated asexually, seeds can be planted outdoors in early spring or in a greenhouse at other times. They should be shaded from direct sun.

HEDERA (*Hedera*) IVY. These start easily from layers or from hardwood or softwood cuttings.

HEMLOCK (*Tsuga*). Seed should be planted in the fall, if possible, or stratified for three months at 40 degrees F. and planted in the spring. The medium should be sterile to avoid damping-off diseases,

and the seedlings must be shaded for at least the first summer. (See Evergreens in Seed chapter.) Dwarf, weeping, and other cultivars can be layered, but this process may take several years. Hardwood or softwood cuttings are sometimes used to propagate hemlocks, but rooting is not easy. Rooting chemicals, mist, and bottom heat are all useful. Cultivars may be side grafted on seedlings.

Hemlock

HIBISCUS (*Hibiscus syriacus*). ROSE OF SHARON, SHRUB ALTHEA. Both this and the Chinese hibiscus (*H. rosa-sinensis*) are easily started from cuttings. Rooting chemicals and mist are both useful. Cultivars of the Chinese hibiscus are sometimes T-budded or grafted on plants grown from cuttings taken from more vigorous varieties. Air layers are sometimes used on those kinds that are difficult to start in other ways.

HOLLY (*Ilex*). Although certain varieties of holly can be grown easily from seed planted soon after ripening, other kinds may take a year or more to germinate. Because the plants are either male or female, and only the females produce berries, asexual propagation is usually preferred so there will be a majority of females. As with all such plants, at least one male in the neighborhood is necessary for pollination.

Although hollies are budded in late summer, and often grafted on potted seedlings in a greenhouse during the dormant season, cuttings are the most popular way of starting them. Cuttings taken in late summer when the wood is nearly mature seem to work well for the evergreen hollies. They should then be treated with rooting chemicals and given mist and relatively high temperatures in a greenhouse. Rooting will be slow, perhaps taking several months. Care must be taken to gradually harden the plants before they are set outside. Since they are acid-loving plants, a mixture of perlite and peat makes a good rooting medium. They should not be given lime or watered with hard water. Deciduous hollies start well from softwood cuttings under mist in early summer.

Home gardeners who want to propagate only a few plants may start them by regular layering, or by air layers.

HONEY LOCUST (*Gleditsia triacanthos*). The common kinds of honey locust can be grown from seeds, which, because of their hard coats, should be soaked in sulfuric acid for an hour, washed thoroughly, and then stratified for three months at 40° F.

The thornless and seedless varieties can be started by hardwood cuttings, which are taken in late winter, callused, and planted out in the spring, or by grafting on seedlings of the thorny type. Seedlings may often be found growing around mature trees.

HONEYSUCKLE BUSH (*Lonicera*). Bush honeysuckle may be started by planting seeds, but the better varieties should be started asexually. The seed should be planted as soon as it is ripe, or stratified for three months or more at 40° F. Hardwood or softwood cuttings root easily. Softwood cuttings do best if taken early, when the plants are making their fastest growth, and are rooted under mist. Both hardwood and softwood cuttings root better if treated with rooting chemicals.

For a limited number of plants, larger shrubs may be dug and split apart, offshoots can be cut away, or they may be layered.

HONEYSUCKLE VINE *(Lonicera).* These are easily layered, which is a useful method to start a limited number of vines. Both hardwood and softwood cuttings root well.

HORSE CHESTNUT. See BUCKEYE

HYDRANGEA *(Hydrangea).* Certain species of hydrangea sucker freely, so plants can easily be split from the parent. They can also be layered or stooled. Cuttings are an easy way to get large numbers of plants quickly. The pink, fall blooming, P.G. *(H. grandiflora),* is often propagated from hardwood cuttings. Both P.G. and other kinds of hydrangea start easily from softwood cuttings taken in early summer and put under mist. They should be removed as soon as they root, before they begin to rot. A perlite-vermiculite medium works well.

HYPERICUM *(Hypericum).* St. Johnswort. Softwood cuttings start easily if taken in late summer from partially mature growth, and rooted under mist or high humidity. Hardwood cuttings may also be used, and the plants can be divided easily in early spring, when they are dormant.

JASMINE *(Jasminum).* Softwood cuttings from partially mature wood root well in a greenhouse, and hardwood cuttings root well. Rooting chemicals and mist are helpful. Small numbers can be started from suckers and layers.

JUNIPER *(Juniperus).* Native kinds can be started from seeds gathered and planted in the fall. Seed germination is spotty, so more should be planted to allow for this, and some

varieties may take up to two years to germinate. Stratification is necessary for seeds not planted as soon as they are ripe, and three to five months at room temperature followed by three additional months at 40 degrees F. is considered ideal.

Named varieties will not come true from seed, and must be started asexually. Creeping and spreading varieties, when layered, often root the same season, and nearly all may be propagated from cuttings, through some varieties root better than others. The upright kinds seem to root with more difficulty, and cuttings are usually taken in late fall and rooted over the winter in a greenhouse. A temperature of at least 60 degrees F. and a high humidity are necessary. A mixture of perlite and peat moss is a good medium. Additional light aids in the rooting.

Commercially, certain named cultivars are grafted on seedling plants. The side veneer graft is the common method. Because considerable care is necessary for a successful union, most home gardeners prefer to increase their plants by cuttings or layers.

KALOPANAX. Seed is likely to take two years to sprout if planted directly outdoors, but it will grow fast once it germinates. If not planted directly, seeds need stratification for six months at a warm temperature, then an additional two to three months at 40 degrees F. Kalopanax may also be propagated by softwood cuttings, root cuttings, or by grafting named varieties on seedlings.

KOELREUTERIA. See GOLDEN-RAIN TREE

LAGERSTROEMIA INDICA. See CRAPE MYRTLE

LARCH *(Larix).* TAMARACK. Seed should be planted in the fall as soon as it is ripe, or stratified for three months at 40 degrees F. Softwood cuttings are difficult to root, but

possible. Special varieties may be grafted, but should be done on seedlings of the same species as the variety to be grafted. Thus, a European larch should not be grafted on a Japanese or American larch seedling.

LIGUSTRUM. See PRIVET

LILAC *(Syringa).* There are many, many varieties of lilacs, most of which propagate easily from layers or division. Many also send up numerous suckers that can be split off.

Most of the late-blooming lilacs including the Prestonia group, such varieties as James MacFarlane, Agnes Smith, and many others, start easily from softwood cuttings taken in early summer. These should be treated with rooting chemicals and kept under mist.

French lilac cuttings and those taken from old-fashioned lilac or similar species when put under mist often lose their leaves before they root. The cuttings should be taken early in the season when they are actively growing and placed in a moist medium. Instead of frequent mist, they should receive only an occasional misting of leaves, and not be kept constantly wet. The medium should be exceptionally well drained, and coarse perlite or sand is often used.

Rooting should take place where there is an abundance of light, but direct sunshine will dry out the plants too rapidly. The plants should be kept covered with a plastic tent to prevent drying out. Temperatures of at least 70 degrees F. are necessary if the rooting is to take place before the leaves drop.

Old-fashioned lilacs, Japanese tree lilacs, and others that are not named cultivars may be propagated by seeds, and they will produce plants much like their parents. Seedling plants can also be used as grafting stock. Plant the seeds as soon as they ripen and dry, in late summer, and germination should take place the following spring. If summer planting is not possible, stratify the seeds over the winter at 40 degrees F. and plant them in the spring. Seedlings of the better varieties often produce plants with outstanding flowers.

Lilac

French lilacs were formerly grafted on green ash, common lilac seedlings, or on rooted cuttings of privet. Most gardeners prefer to have lilacs growing on their own roots, however. Lilacs grafted on privet sometimes sucker badly and are often not hardy in northern areas. If you choose to graft lilacs, it is possible to do so outdoors, but most propagators prefer to pot the seedlings they use for rootstock in the fall, store them over the winter, and graft them in early spring in a greenhouse. Commercially, side veneer grafts are often used, but amateur grafters often have better success with cleft grafts. Grafting also makes possible the novelty of several colors on the same bush.

Lilacs may also be bud grafted in summer. Both budding and grafting should be done as low on the stock as possible so that when the graft is planted in its permanent location, it will be below ground level, which will encourage the scion to grow its own roots.

Lilacs growing on their own roots can be propagated by root cuttings, though this method is not common.

LINDEN *(Tilia)*. The seeds should be planted as soon after they ripen as possible. Older seeds develop a hard coat, and must be soaked in sulfuric acid, rinsed thoroughly, and stratified until spring in moist peat moss for satisfactory germination; or they may be stratified for four months at a warm temperature and four additional months at 40 degrees F.

Named varieties are usually budded or grafted on seedlings of the same species. Suckers or offshoots growing around tree stumps can often be dug off and planted, and young trees may have their lower branches layered. Seedlings of the big leaf linden (basswood) are often found growing along shady northern country roads, and these may be dug and transplanted in late fall or early spring.

LIQUIDAMBAR *(Liquidambar styraciflua)*. Seed should be planted as soon as ripe or stratified for two months at 40° F. Even so, they may not germinate for two years. Softwood cuttings are possible, and named cultivars are grafted or budded on seedlings.

LIRIODENDRON TULIPIFERA. See TULIP TREE

LOCUST, BLACK *(Robinia pseudoacacia)*. The seeds should be soaked in hot water overnight, or for an hour in sulfuric acid, and washed in water before planting to soften the hard seed coat. Certain varieties sucker freely, and these can be dug up and transplanted.

LONICERA. See HONEYSUCKLE

LOTUS *(Lotus)*. These are of the pea family and not to be confused with the water lotus *(Nelumbium)*. Plants may be divided. Cuttings taken just after flowering root easily, and seeds are also used.

MADRONE, PACIFIC *(Arbutus)*. Cuttings taken from late fall through the winter root well, if treated with rooting chemicals. A peat-sand mixture is a good medium.

Choice cultivars are propagated by grafting. Seeds are also used, and should be stratified for three months. Seedlings are best transplanted early to pots, because the larger plants do not transplant well. Propagation can also be done by layering if only a few plants are wanted.

MAGNOLIA *(Magnolia)*. Magnolia seeds should be planted as soon as they are ripe, or stratified for four months. The seedlings grow rapidly, and should be transplanted while still small to pots or other containers, since the trees are difficult to dig because of their extensive root system.

Some varieties can be started easily from cuttings taken from young trees. These should be of partially matured wood, taken in summer, and treated with a rooting chemical. They root well in sand, especially under mist. Rooting them in small pots prevents transplanting difficulties later.

Named cultivars may be grafted with a side or veneer graft in the greenhouse during the winter. The seedlings of the cucumber tree *(M. acuminata)* are often used as rootstock for the magnolia hybrids, and *M. grandiflora* is grafted on its own seedlings.

Small numbers of magnolias may be started by layering young trees. These make take a year or more to root.

MAHONIA *(Mahonia)*. Seeds should be planted before drying out, but still may take two years to germinate. Stratify, if not planted right away, for three months at 40 degrees F. Some varieties root fairly well from softwood cuttings taken in summer, treated with rooting compounds, and placed under mist. The plants also root well from hardwood cuttings, or they may be layered or divided.

MALUS. See APPLE (Fruit section) or CRABAPPLE

MAPLE *(Acer).* The many maple species are easily grown from seed. Seeds of most varieties germinate best when planted immediately, and they should not be allowed to dry out. Red maple *(A. rubrum)* and silver maple *(A. saccharinum)* ripen their seeds in early summer, and these should be planted at once. Other varieties ripen their seeds in the fall, and if not planted immediately, they should be stratified for three months.

Maple

Named varieties of maples, or especially good selections of native trees, such as those that turn a nice color in the fall, can be propagated by grafting. Grafting or T-budding must be done only on seedlings of a closely related variety. Thus, cultivars of Norway maple, such as Crimson King, should be grafted on Norway maple seedlings. Those of Japanese maples such as Burgundy Lace should be grafted on Japanese maple seedlings, and so forth.

Softwood cuttings should be taken in late spring or early summer when the trees are growing rapidly. Those from young trees root best, and it helps to use a rooting compound. Even under the best conditions, cuttings do not root easily, but those taken from trees grown from cuttings sometimes root fairly well. A warm greenhouse and a mist system are almost necessities for successful rooting of maples. The rooted cuttings may need to be protected over their first winter.

Young trees can also be layered or stooled successfully, and if it is done early in the spring, rooting may take place the first season. Although most maple cultivars are grafted, if they can be started on their own roots they usually live longer than grafted trees, because in time the graft may separate. This is often an especially serious problem in areas that experience low winter temperatures.

MELIA *(Melia azedarach).* CHINA-BERRY. The seeds of this popular southern tree, a native of the Himalayas, often self-sow. Seedlings can usually be dug up from around larger trees. Softwood cuttings taken in early summer root easily.

METASEQUOIA *(Metasequoia glyptostroboides),* DAWN REDWOOD. Hardwood cuttings root easily, and softwood cuttings under mist do equally well. The use of rooting chemicals helps greatly. Seedlings vary, but the seeds germinate and grow so well that they have been reproducing the tree for over fifty million years. This fact alone makes it an especially interesting tree to have on one's property. Plant the seed as soon as it is ripe.

MOCK ORANGE *(Philadelphus).* Both hardwood and softwood cuttings root quickly and with little care. The use of rooting chemicals helps, and it is advisable to use mist on the softwood cuttings. The plants can also be divided, and suckers coming from the roots can be dug off and transplanted in spring. Quite a number of plants can be started by layering the low-growing branches. Stooling the plant also works well.

MYRICA *(Myrica).* Both the Pacific wax myrtle *(M. californica)* and the bayberry *(M. pennsylvanica)* are usually propagated from seeds. The wax around the seed should be removed with hot water and the seeds planted at once before they dry out. Otherwise, they should be stratified for three months at 40 degrees F.

Myrica may also be layered and started from cuttings taken in summer and put under mist. The sexes are separate in this species, and since only the female plants have berries, asexual propagation is the only certain way of getting some of each sex.

NANDINA *(Nandina domestica).* HEAVENLY BAMBOO. If the seeds are gathered in the fall and stored at a low temperature until late the following summer, they should germinate. Seedlings grow slowly. The plants may also be layered, or root suckers may be dug out and transplanted.

NERIUM. See OLEANDER

NINEBARK *(Physocarpus).* The common varieties can be started from seeds, as soon as they are ripe, but the dwarfs and those with yellow foliage should be started asexually. Hardwood cuttings root easily, as do softwood cuttings, especially under mist. Rooting chemicals are helpful. Larger plants may be divided in early spring.

NORFOLK ISLAND PINE *(Araucaria heterophylla).* These ornamental trees that are often grown in pots in the north can be started from seeds or by cuttings. Cuttings taken from side branches tend to grow horizontally, while those from terminal branches grow upright.

OAK *(Quercus).* The oak family is a large one, and there is wide variation in the time it takes for a seed to start. Acorns from white oak should be planted as soon as they drop. Those from black oak should be stratified for a month or more at 35 degrees F. before planting. Most other kinds can be kept for a short time without stratification, but do not allow them to dry out. If kept for several months, they too should be stratified at 35 degrees F. for a month or more.

Named varieties of oaks are best grafted on seedlings of the same type. Thus, scarlet oak cultivars such as Splendens, should be grafted only on scarlet oak seedlings, and not on red or white oak seedlings. Starting oaks from cuttings is very difficult.

OLEANDER *(Nerium)*. Softwood cuttings taken in the summer root well in water or in rooting medium under mist. The top should be pinched off soon after rooting to encourage a bushy growth habit.

Oleander may also be started by air layers and by seeds. Seeds should be rubbed to take off their fuzzy coating and planted in flats in a greenhouse.

PAEONIA SUFFRUTICOSA. See PEONY, TREE

PALMS. There are thousands of palm varieties, and most are propagated by seed. In very warm areas they may be planted outdoors, but in most places it is better to plant them in flats containing a mixture of half perlite and half peat moss, in a warm greenhouse. It may take several months, or even a year or two, for some varieties to germinate. The seedling plants should be carefully watched and sprayed with a fungicide if any molds develop.

Some palms may be propagated from offshoots or suckers that appear at the base of the plant.

PARTHENOCISSUS *(Parthenocissus)*. The Boston ivy *(P. tricuspidata)* and the Virginia creeper *(P. quinquefolia)* are woody vines that can be started easily from layers, cuttings, or seeds. Softwood cuttings started in the summer root well under mist, even with no rooting chemicals. Hardwood cuttings may be planted directly in the ground in early spring, and a great many will root.

Seed should be planted as soon as it is ripe in the fall, or else stratified for three months and planted in the spring. Because layered vines root so rapidly, a considerable num-ber of plants can be started by compound layers, or stools each year.

PEAR, BRADFORD *(Pyrus calleryana Bradford)*. This cultivar of the Chinese Callery pear is ornamental, with large white flowers and very small fruits. Because it is resistant to fire blight, seedlings from it are often used as rootstock for fruiting varieties of pears. Softwood cuttings that have been treated with rooting compounds can be rooted under mist, but the Bradford pear is usually T-budded on seedling roots when propagated commercially.

PEONY, TREE *(Paeonia suffruticosa)*. The common tree or moutan peony is commonly started from seeds, planted as soon as they are ripe. It is from China, and hundreds of named varieties have originated from it. Most of these cultivars are started by dividing the plant. Layering is possible, too, but rooting is slow.

Named varieties are often grafted on tree peony seedlings. These should be grafted close to the ground so that when transplanted, the union can be set below the surface of the soil, enabling the cultivar to grow roots of its own above the graft.

PHILADELPHUS. See MOCK ORANGE

PHYSOCARPUS. See NINEBARK

PICEA. See SPRUCE

PIERIS ANDROMEDA. Seeds can be sown as soon as ripe or stored in airtight containers in a cool place for several months. *P. floribunda*, the mountain andromeda, can be rooted from cuttings, but only with difficulty. Both hard and softwood cuttings of Japanese andromeda *(P. japonica)* root quite easily.

PINE *(Pinus)*. Most varieties of pine are propagated by seeds, although certain dwarf kinds and cultivars of the weeping, upright, and oth-ers are grafted on seedling pines of the same species. Callused cuttings are used also, and a few, such as the mugho, can be started from softwood cuttings, treated with rooting compounds, and placed under mist. Certain spreading pines can be layered, but it may take several years for them to root.

Fall is the best time to plant seeds. (See Evergreen seed in Chapter II.) Seeds that cannot be planted at once should be stratified at 35 degrees F. for two months before planting. Some varieties remain viable for several years in a sealed container, if kept cool and dry. It is best to plant them in sterilized soil or an artificial soil medium, so damping-off diseases will be less of a problem.

Named cultivars of pines are usually grafted in a greenhouse. Scions of the wanted varieties are side or veneer grafted on potted seedlings of the same species. (See Veneer Grafting.)

PISTACIA, CHINESE *(Pistacia chinensis)* PISTACHIO. The sexes are separate on these trees, and if they are to be grown for nuts, the usual method is to plant a ratio of one male to six females. Planting seeds is the usual method of propagating trees grown for ornamental purposes. They should be separated from the pulp soon after it ripens and stratified for two or three months at 40 degrees F. before planting, for better germination. For growing nuts, the best varieties are often grafted on potted seedlings of *P. chinensis*.

PITTOSPORUM *(Pittosporum)*. The seeds germinate easily, especially if soaked for a few seconds in hot water before planting. Cuttings root well when they are taken from partially mature wood, treated with rooting chemicals, and placed under mist.

PLANE TREE *(Platanus)* SYCAMORE. The plane tree is usually propagated by seeds, which are best gathered from the tree in late winter, taken from their pods, and planted as soon as possible. If the seeds are gathered or purchased in the fall, they should be stratified for two months at 40° F. before planting. Softwood or hardwood cuttings may also be used to propagate the plane tree.

PLUM, FLOWERING. See CHERRY, FLOWERING

PLUMBAGO *(Plumbago)*. These semi-tropical plants are sometimes grown in pots or tubs in the North. They can be started from divisions or cuttings, or they can be grown from seeds sown in the winter.

PODOCARPUS *(Podocarpus)*. These grow readily from cuttings taken in late summer or early fall and rooted in the greenhouse with rooting chemicals. They seed themselves in their native tropics.

POINSETTIA *(Euphorbia pulcherrima)*. Cuttings can be rooted nearly any time the plants are growing actively. A rooting chemical helps, as does mist. Larger plants can be divided, and they can also be layered. Both cuttings and layers are best done in pots, so the roots will be disturbed as little as possible by transplanting. The tops of new plants must be pinched to get a bushy effect, and the day length and temperature should be controlled to get them to bloom at Christmas.

POPLAR *(Populus)* ASPEN, BALM-OF-GILEAD, COTTONWOOD, POPPLE. Most poplars sucker freely, and new plants can be easily started by digging the suckers in the spring. Freshly gathered seeds also grow easily, as do both hardwood and softwood cuttings. No special treatment is usually necessary. Root cuttings and layers are also methods of propagating this vigorous-growing species.

POTENTILLA *(Potentilla)* CINQUE-FOIL. This low-growing shrub, which is useful because of its compactness, hardiness, and long-blooming characteristics, is easily propagated by softwood cuttings under mist, with the use of a rooting chemical. They may also be propagated by layering. Large shrubs can be divided when dormant. Seeds are planted mostly to develop new varieties.

PRIVET *(Ligustrum)*. Since hardwood cuttings root easily, they are a common way of starting privet. Large numbers of plants can be started from softwood cuttings in early summer when hedges are being sheared. The use of a rooting chemical and mist help ensure quicker and better rooting. Privet can also be started from seed which should be stratified for two or three months at 40 degrees F. before planting.

PRUNUS. See CHERRY, FLOWERING

PSEUDOTSUGA. See DOUGLAS FIR

PYRACANTHA *(Pyracantha)* FIRE-THORN. Softwood cuttings, taken in early summer and treated with rooting chemicals, root well under mist. Semi-hardwood cuttings taken in late summer can be rooted in a greenhouse in the fall. Large cuttings, one to two feet long, produce good-sized plants quickly.

PYRUS CALLERYANA. See PEAR, BRADFORD

QUERCUS. See OAK

QUILLAJA SAPONARIA *(Quillaja saponaria)* SOAPBARK TREE. This tropical tree is grown in the warmer parts of California, and is started mostly by softwood cuttings under mist, using a rooting chemical.

QUINCE, FLOWERING *(Chaenomeles)*. Hardwood cuttings root well, as do softwood cuttings that are treated with a rooting chemical, and placed under mist. Plants may be layered, and suckers dug from around older plants. Clumps can be divided, and they may be propagated easily by stooling. They may also be started from seed, which should be cleaned from the fruit before planting in the fall, or stratified for at least two months at 40 degrees F., and planted in the spring.

RASPBERRY, FLOWERING *(Rubus odoratus)*. Flowering raspberries are easily propagated by division, layers, and softwood, hardwood, and root cuttings. Seeds should be planted as soon as ripe or stratified at a warm temperature for three months, then at 40 degrees F. for three months.

REDBUD *(Cercis)*. Lower branches may be layered. Cuttings of new growth taken in early summer root well when treated with rooting chemicals and placed under mist. Seeds give fairly good results, and must either be planted in the fall, soon after ripening, or soaked for an hour in sulfuric acid, rinsed well, and stratified for ninety days at 40 degrees F. Named cultivars are grafted or T-budded on seedlings or sucker plants.

REDWOOD *(Sequoia)*. Fall planting of the seed as soon as it is mature and comes out of the cone easily is the easiest method of propagation. Seed to be planted at other times should be dried thoroughly, and stratified for sixty days at 40 degrees F. Seed beds should be kept shaded. Cuttings taken from shoots that appear on the burls are sometimes rooted.

REDWOOD, DAWN. See META-SEQUOIA

RHAMNUS. See BUCKTHORN

RHODODENDRON *(Rhododendron)*. There are thousands of varieties in this family, which also includes azaleas (which see). Seeds are used for starting the varieties that reproduce more or less true from this method. They are also used for starting rootstocks upon which the named varieties are grafted. The seed should be collected after it ripens, placed in an upright container, and stored in a cool dry place until early spring when it is planted in the greenhouse.

Because the planting medium should be acid, a mixture of vermiculite and peat or sphagnum moss is good. Water with rain water, or other water that is not alkaline and is rich in calcium. Keep the flats out of direct sunlight, and hold the temperature fairly constant—between 60 and 70 degrees F.

Since the seedlings grow slowly, it may be several months before it is safe to transplant them. They should be kept in a sheltered environment until the following spring, when they may be planted out or potted for grafting.

If you want only a few plants, layering is a simple way to propagate. Grafting of rhododendrons is most successful if done in a greenhouse, using a side veneer graft. The newly grafted plants should be kept in an area of high humidity at a warm temperature, until the graft union is established. Gradually, then, they may be moved to a cooler, less humid environment.

Most broadleaf evergreen cultivars, however, are started from cuttings. Fall is the usual time to take these, and they are then rooted in a warm greenhouse over the winter. Rooting chemicals and bottom heat are most beneficial, and an acid medium is necessary. Regular sprays of fungicides may be needed to control the diseases that often develop in warm, humid conditions. Plants started from cuttings and layers are better than grafted ones because there is not the problem of having undesirable plants growing from root suckers.

RHUS. See SUMAC

RIBES *(Ribes)* FLOWERING CURRANT, ALPINE CURRANT. These ornamental shrubs can be started from sucker plants, divisions, and hardwood and softwood cuttings. Seeds should be stratified for three months at 40 degrees F. See Fruit

section for edible currants and gooseberries.

ROBINIA PSEUDOACACIA. See LOCUST, BLACK

ROSE *(Rosa)* TEA, FLORIBUNDA, MULTIFLORA, and FLORABUNDA ROSES. In the desire to get the best possible blooms, hybridizers have so weakened these popular roses that they now have little vigor. Although they can be started from cuttings, layers, and stools, the new plants usually grow poorly, and have few blooms. For this reason they are ordinarily propagated by bud grafting them on hardy, vigorous growing rose seedlings. The most commonly used stock throughout the country is the thornless *R. multiflora.* It suckers less than many common roses and is easily grown from seed planted in the fall soon after the hips ripen, or it can be stratified at 40 degrees F. for six weeks, and planted in the spring. In warm areas, where soil diseases and soil nematodes are a common problem, more resistant varieties are used.

Rose

Roses grown in greenhouses exclusively for producing flowers for florists are also grafted on disease-resistant stocks, such as *R. odorata,* which is usually propagated from cuttings. Though it is not a hardy stock, it has proven to be well suited for greenhouse conditions.

Hardy Rambling Roses

These can be propagated by layering. By using serpentine layers in early spring, a large number may be started. They also grow well from softwood cuttings under mist.

Climbing Roses

Climbing roses can be started by either hardwood or softwood cuttings, as can the miniature roses often grown as pot plants.

Tree Roses

These are quite difficult to grow and their propagation is probably best left to the experts. Usually an upright type of rose such as a shrub or briar rose is grafted on a seedling of a multiflora rose. These are allowed to grow for a year, then grafted a few feet above the ground with a tea rose variety. Since both the rootstock and the interstem tend to sucker, it is necessary to remove all unwanted growth frequently.

Shrub Roses, Specie Roses, Wild Roses

The hardy roses grow with little care and can be propagated by division, taking offshoots or suckers, layering, hardwood and softwood cuttings, stools, and seeds. Special varieties such as Harrison's Yellow and rugosa hybrids do not come true from seed and must be started asexually.

ROSE OF SHARON. See HIBISCUS

RUSSIAN OLIVE. See ELAEAGNUS

ST. JOHNSWORT. See HYPERICUM

SALIX. See WILLOW

SAMBUCUS. See ELDER

SEQUOIA. See REDWOOD

SERVICEBERRY *(Amelanchier)* SHADBUSH, and others. This group of berry-producing plants encompasses a large variety ranging from small shrubs to small trees. Seed planted as soon as it ripens germinates well; otherwise it should be stratified over the winter and planted in the spring. Softwood cuttings root fairly well with a low level of mist and the use of rooting chemicals. Root cuttings taken in early spring grow well. Larger shrubs may be divided, and low-growing branches may be layered, but the different varieties vary widely in how fast they root. Root cuttings taken in early spring grow well, also.

SILK-TREE. See ALBIZZIA

SMOKETREE (Cotinus coggygria). Although it can be propagated from seeds, many of the seedlings produce plants without showy blooms. Softwood cuttings taken in late spring root well if rooting chemicals and mist are used. These should be taken only from plants that produce the best blooms or the best colored leaves.

SORBUS (Sorbus) MOUNTAIN ASH, ROWAN TREE, MOOSE-MISSI. Seeds require a rather critical after-ripening period. Since birds often pick the berries early, it may be necessary to protect them until the seed has ripened. Then clean the seeds from the berries and stratify them in sphagnum moss for several weeks in a warm greenhouse before planting. They may then be planted outside, or stratified at 40 degrees F. for several months and planted in the spring.

It is very difficult to propagate mountain ash by either layering or stem cuttings, though root cuttings may succeed. Named cultivars are grafted or T-budded on seedling rootstocks.

SPIREA (Spiraea). Softwood cuttings taken in early summer root easily under mist. A rooting chemical is helpful. Hardwood cuttings taken in winter and planted in spring also do well. Small numbers of plants may be started from suckers, layers, and divisions.

SPRUCE (Picea). Seed is the easiest and best method to start the forest varieties. As soon as it is ripe, the seed can be easily shaken loose from the cone and should be planted at once if possible. Follow the directions given in Evergreen Seeds on page 41. Seedlings of trees such as the Colorado blue spruce do not always resemble their parents, and a high percentage of them may be various shades of green, rather than a rich blue.

Dwarf spruce and cultivars with outstanding color are usually started from grafts. Grafting is done on potted seedlings of the same variety in a greenhouse using the side veneer graft.

Cuttings are difficult to root at best, but when they are taken in late winter and rooted in a greenhouse with bottom heat, extra light, and rooting chemicals, results are sometimes quite good.

Some of the spreading types of dwarf spruce may be propagated by layering the plants, but they may take several years to root.

STAR-JASMINE (Trachelospermum jasminoides). It is easy to propagate jasmine by layers. Softwood cuttings taken in early spring with a bit of old wood attached to the base root well under mist if a rooting chemical is used.

SUMAC (Rhus). The varieties that sucker freely can be dug up and transplanted. Cuttings taken in midsummer, when treated with a rooting chemical, root well if placed under mist. Seeds should be gathered as soon as ripe, soaked in sulfuric acid for one hour, washed, and planted immediately for the fastest germination. Otherwise they should be stratified five months in a warm greenhouse, followed by three months at 40 degrees F. Some shrubs can be divided.

SYCAMORE. See PLANE TREE

SYRINGA. See LILAC

TAMARACK. See LARCH

TAMARISK. (Tamarix). Softwood cuttings taken in early summer root well under mist if a rooting compound is used. The more common method of propagation is to root hardwood cuttings that are about twelve inches in length.

TAXODIUM (Taxodium) BALD-CYPRESS. Seed should be stratified for a month or more at 40 degrees F. and planted in the spring. Better varieties can be grafted on seedlings of the same species, and hardwood cuttings are possible.

TAXUS. See YEW

THUJA. See ARBORVITAE

TILIA. See LINDEN

TRACHELOSPERMUM. See STAR-JASMINE

TRUMPET CREEPER (Campsis). These may be propagated by layering, but hardwood, softwood, and root cuttings also start well. Seeds may also be used, requiring stratification for two months at 40° F.

TSUGA. See HEMLOCK

TULIP TREE (Liriodendron tulipifera). YELLOW POPLAR. Fall planting of seeds is best, or stratify the seeds for sixty days at 40 degrees F. and plant in the spring in flats or outdoors. Root cuttings work well, as do softwood cuttings that are taken in early summer, treated with rooting chemicals, and put under mist. Outstanding varieties may be grafted on seedlings. Both seeds and cuttings are best started in pots, because even small trees are difficult to dig and transplant.

ULMUS. See ELM

VIBURNUM (Viburnum). There is a great variety of shrubs in this family, and many are useful in the landscape and for attracting bird life. Most plants layer easily and can be divided, and some send up numerous suckers.

To start a lot of plants, softwood cuttings are a good method. They root quite easily when treated with a rooting chemical and placed under mist. The cuttings are best taken early in the season after they have made a growth of six or seven inches.

Cultivars of viburnum may be grafted on layered plants, rooted cuttings, or seedlings of the Arrowwood *(V. denatum)* or of the Wayfaring tree *(V. lantana)*. Certain larger-growing viburnums are occasionally grafted on a dwarf-growing variety in order to get a smaller plant.

If the plant to be propagated is not a named cultivar, it can usually be started easily from seed. Seed can be collected or bought, but the best germination results when it is planted as soon after harvesting as possible. The necessary stratification takes place naturally in the soil, and the plants begin to grow the following spring. If seeds must be held in storage, they should be stratified at 40 degrees F. for two to four months.

VINCA *(Vinca)* MYRTLE, PERI-WINKLE. This viney ground cover layers naturally, and digging and dividing the clumps usually provides a large number of plants. Potted plants may be obtained by sinking pots filled with soil around an established myrtle plant and training the vines to root in them. Larger quantities of plants may be started from cuttings taken in late summer, and rooted under mist.

VIRGINIA CREEPER. See PARTHE-NOCISSUS

WAX MYRTLE. See MYRICA

WEIGELA *(Weigela)*. This plant can be easily started from softwood cuttings taken in early summer and rooted under mist. Hardwood cuttings planted in early spring root well, too. Use of a rooting chemical helps with both methods.

WILLOW *(Salix)*. Willows can be started from seed, layers, and root cuttings, but because they grow extremely easily from both hardwood and softwood cuttings, that method is usually employed.

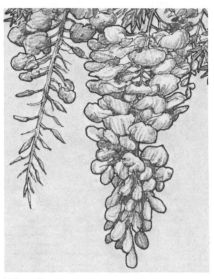
Wisteria

WISTERIA *(Wisteria)*. The easiest way to start this plant in quantity is by softwood cuttings. Treat them with a rooting chemical and place them under mist. Smaller numbers can be started by layering, or by digging out sucker plants. Special cultivars are grafted on more ordinary types. Hardwood cuttings of certain varieties root successfully.

Whatever method is used, it is usually a good idea to start the young plants in small pots, to minimize the losses that often come when the bare-rooted vines are transplanted.

WITCH HAZEL *(Hamamelis)*. Seed is the usual method, and fall is the best time for planting. It may take two years to germinate, however. Germination may be speeded by stratifying the seed five months at room temperature, followed by three months at 40 degrees F. Plants may also be layered. Often, wild plants can be dug up where they are growing abundantly. Some varieties, such as Chinese witch hazel, are grafted on seedlings of the common witch hazel.

WOODBINE. See PARTHENOCISSUS

XANTHOCERAS *(Xanthoceras sorbifolia)*. YELLOWHORN. This plant can be started from seeds that are stratified at 40 degrees F. over the winter and sown in the spring, or

by root cuttings dug in the fall, cut into pieces three inches long, and planted in later winter. Both seeds and root cuttings should be started in pots, because transplanting is difficult otherwise.

YELLOW POPLAR. See TULIP TREE

YEW *(Taxus)*. The Japanese upright growing yew *(T. capitata)* is often started from seed, because it comes fairly true to type. Seeds should be stratified for five months at room temperature, followed by three months at 40 degrees F.

All other yews are usually grown from cuttings. Semi-hardwood cuttings, taken in late fall and rooted under mist in a warm greenhouse or hotbed, root well. They should be rather large, twelve to twenty-four inches long, and include some wood that is more than a year old. Cuttings taken from the tops of upright growing yews tend to produce upright growing plants, while those taken from the sides develop into more spreading shapes. Obviously, yews grown from cuttings don't always resemble their parent.

Although the berry-producing female plants have a certain attractiveness, the male plants root heavier and grow faster, so therefore are most often propagated. Cuttings should be removed from the mist as soon as the rooting has taken place, but they should not be transplanted until they have developed a heavy root system. Even after they have been planted in the ground, constant pinching and shearing will be necessary to produce a compact plant.

ZELKOVA *(Zelkova)*. These trees have been closely watched to see if they could fill the void left by the American elms that have been lost to Dutch elm disease. Although they are not nearly as hardy as the elm, they have proved to be worthy trees where they can be grown. They are best propagated by seeds that have been stratified for about three months at 40 degrees F.

HERBACEOUS PLANTS

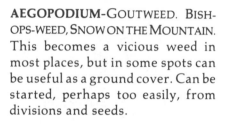

You will find it necessary to read carefully the first section of this book, especially the chapters on Home Nursery, Seeds, Cuttings and Division, before trying to use these instructions.

You will note that many of the perennial plants listed can be started from seed, and although most will not bloom until the second year after planting, many will bloom the first year if they are planted in late winter in the house or greenhouse.

ACHILLEA-YARROW. Hardy perennial. Divide clumps, or take cuttings in mid-summer to propagate best varieties. Seeds germinate in ten to fifteen days at 70 degrees F. but may produce inferior plants.

ACHIMENES. Tender perennial, usually grown as a houseplant. Divide the very small rhizomes in early spring, and plant about one-half inch deep in pots indoors. Also take softwood cuttings in spring.

ACONITUM-MONKSHOOD. Hardy perennial. Plants are easily divided, and one large clump makes many new plants. Seed best kept in the refrigerator for six weeks before planting. Germinates in fifteen to twenty days.

ACROCLINIUM (*Helipterum*)-SUN-RAYS. Annual. Seeds germinate in a week.

AEGOPODIUM-GOUTWEED. BISH-OPS-WEED, SNOW ON THE MOUNTAIN. This becomes a vicious weed in most places, but in some spots can be useful as a ground cover. Can be started, perhaps too easily, from divisions and seeds.

AETHIONEMA-STONE-CRESS. Perennial rock plant. Can be started from cuttings in early summer, by division, or from seeds sown in spring.

AFRICAN VIOLET. See SAINT-PAULIA

AGAVE. Perennial succulent plant. Offshoots can be cut or broken off and planted, or suckers dug. Sow seeds in sandy soil in a warm place. Germinates in thirty days.

AGERATUM HOUSTONIANUM-AGERATUM. Annual. Start seeds inside for early blooms (70 to 80 degrees F.). Cuttings from established plants root easily.

AJUGA-BUGLE. Hardy perennial ground cover. Can be started easily from divisions anytime. Seeds grow easily, too.

ALLIUM. ORNAMENTAL ONIONS, including CHIVES. Hardy perennial. Most species grow well from seed, or by division of the bulbs.

ALOE. Tender perennial. Medicinal herb. Separate offshoots, or divide the plants. Seeds planted in sandy

soil germinate in twenty to twenty-five days at 70 degrees F. Cuttings root well if allowed to dry a few hours before placing them in sand or vermiculite.

ALTHAEA ROSEA-HOLLYHOCK. Hardy biennial that sometimes lives three to four years. Seeds grow easily, and often self-sow. Germinates in ten days, and blooms the second year. Some varieties, such as Indian Spring, are annuals, and may be planted for blooms the same year.

ALYSSUM. GOLDEN TUFT. A fairly hardy perennial. Should not be confused with annual sweet alyssum (*Lobularia*). Divide in the spring, or start by softwood cuttings after the new growth is a few inches tall. Seed can be planted as soon as it is ripe, or in summer for bloom the next year. It germinates in three or four weeks at 70 to 80 degrees F.

AMARANTHUS CAUDAUTUS. Annual. Plant seeds inside for earlier results, and give extra light. Germinates in fifteen to twenty-one days at 70 to 85 degrees F.

AMARYLLIS. See HIPPEASTRUM

AMARYLLIS BELLADONNA. Tender perennial. Grown as a pot plant in the North, outdoors in warm areas. Separate the bulbs or take bulb cuttings to start new plants.

ANCHUSA-BUGLOSS. Annual and perennial. Start seeds early inside. Germinates in two to three weeks at 70 to 85 degrees F. Propagate outstanding varieties by root cuttings or divisions of clumps.

ANEMONE. Perennial. Most anemones can be started from seed, sown in the spring, which germinates in fifteen days, or from division of the roots in early spring. The hardy perennial anemones include the Japanese, which can be propagated either by division or from two-inch pieces of root which, when planted in sandy soil, will form new plants.

The poppy anemone (A. coronaria), a tender perennial, is the most showy and the one sold by florists. Seeds are started at 70 degrees, and take at least five weeks to germinate. Higher temperatures may injure them. The tubers are dug and dried after blooming and then replanted. They can be divided at that time. The tubers can also be bought from bulb dealers.

ANTHEMIS TINCTORIA-MARGUERITE. YELLOW DAISY. Hardy perennial. Dividing the plants is the easiest way to start a lot of plants. Cuttings taken in mid-summer are also possible. Seed germinates in from one to two weeks. Self-sows rapidly.

ANTHURIUM ANDRAEANUM-Tender perennial. Take offshoots, with attached roots, in later winter, and plant in pots kept in a closed case for several weeks. Sow seeds in a sand-sphagnum moss mix, and keep covered with glass or clear plastic.

ANTIRRHINUM MAJUS-SNAP-DRAGON. Tender perennial, usually grown as an annual. Seeds should be started inside for early blooms, germinate in two to three weeks at 55 degrees F. or higher. Can also be started from cuttings.

AQUILEGIA-COLUMBINE. Hardy perennial that grows easily from seed sown in spring or mid-summer. Germinates in two to four weeks. Older clumps can be divided but the resulting plants are less vigorous and shorter-lived than seedlings.

ARABIS-ROCK-CRESS. Hardy perennial. Plants can be divided in early spring or fall, and cuttings from new growth root easily. Seeds germinate at 70 degrees F. in two to four weeks. Additional lighting speeds germination and growth.

ARCTOTIS STOECHADIFOLIA-AFRICAN DAISY. Annual. Sow seeds inside for early blooming. Germinate in two or three weeks at 70 degrees F.

ARMERIA - THRIFT. Hardy perennial. Evergreen. Best method of propagation is to divide clumps in spring or late summer. Seeds will germinate in three or four weeks at 70 degrees F.

ARTEMISIA-WORMWOOD, SILVERMOUND, DUSTY MILLER, etc. These hardy perennials, with ornamental foliage that is often fragrant, can be increased rapidly by divisions or cuttings, and can also be grown from seed.

ARTICHOKE, JERUSALEM (Helianthus tuberosus). Hardy perennial related to the sunflower that produces edible tubers. Divide the tubers in spring or fall. It may be started also from soft cuttings in early summer. In some areas it spreads rapidly naturally and becomes weedy.

ARUNCUS SYLVESTER - GOATS BEARD. Hardy perennial. Clumps may be divided in the spring. Seed grows easily, but the male plants produce no seed.

ASCELPIAS TUBEROSA - BUTTERFLY WEED. Hardy perennial. Division of the plant is difficult because of its long taproots. Seed germinates in three or four weeks at 70 to 85 degrees F. and is the most sure way to propagate. A. curassavica may also be propagated by soft cuttings.

ASPARAGUS ASPARAGOIDES - FLORIST ASPARAGUS. Tender perennial. Plant seeds as soon as they ripen. They germinate in about three weeks at 70 to 80 degrees F. Clumps can be divided in early spring, and cuttings of new side growth may be rooted.

ASPARAGUS OFFICINALIS - The eating varieties are easily grown from seed, which germinate in one to three weeks. Transplant the following spring to the spot it should grow permanently. The crowns may be divided in early spring or fall, but this disturbs production for a few years.

ASTER-MICHAELMAS DAISY. Hardy perennial. Divide clumps in fall or early spring. Seeds germinate in two or three weeks at 70 degrees F. and usually self-sow.

ASTER CALLISTEPHUS. Annual. Seed should be sown inside for earliest blooms. Germinates in about eight days at 70 degrees F.

ASTILBE - SPIRAEA. Divide the clumps in early spring.

AUBRIETA DELTOIDEA - PURPLE ROCK-CRESS. Hardy perennial. Cuttings taken directly after blooming root well. Seeds germinate in about fifteen days at 55 degrees F. May be divided in spring or fall, but not easily.

AUTUMN CROCUS. See COLCHICUM

BALLOON FLOWER. See PLATYCODON

BAPTISIA - FALSE INDIGO. Hardy perennial. Clumps are not easy to divide, but, with care, it is possible. Seeds germinate unevenly and should be planted outdoors as soon as they are ripe.

BEE BALM. See MONARDA

BEGONIA. Tender perennials. The fibrous rooted and wax begonias may be propagated by stem cuttings, leaf cuttings, division, or by seed. The seeds are tiny and need extra care in planting and after care. An artificial soil mix is best and the seeds should be very lightly covered with it, or left uncovered, the flat covered with a sheet of clear plastic. They should germinate in two to three weeks at 70 degrees F. if given extra light. Some varieties do not come true from seed and must be propagated by asexual methods.

Tuberous begonias can be started by seed, but the tiny seedlings must be treated and transplanted with special care. The large bulb-like roots of both tuberous and Rex begonias may be cut apart to start new plants, but make sure there is a sprout on each portion. Both can be started from stem or leaf cuttings, which should be taken early in the season when they root most easily.

BELLIS - ENGLISH DAISY. These are hardy perennials, but are short-lived so they are often treated as annuals or biennials. Usually they self-sow, and last for many years in this way. Clumps of the best varieties should be divided every spring. Seed starts in a week or two at 70 degrees F.

BLEEDING HEART. See DICENTRA

BOCCONIA CORDATA (*Macleaya*) PLUME POPPY. Division is the easiest method of propagation. In fact, they should be separated each spring, or they may spread throughout the garden. Root cuttings are also possible, and seeds sprout and grow readily.

BOLTONIA - FALSE STARWORT. Hardy perennial. This aster-like flower can be started easily by division in spring or fall, which is usually the best way to propagate it. The seeds sprout in two to three weeks at 70 degrees F.

BROWALLIA - AMETHYST FLOWER. Tender perennial, usually grown as an annual, and sometimes used as a winter indoor plant. Cuttings root easily, but it is usually started

from seeds which germinate in about two weeks at 70 degrees F.

BUGBANE. See CIMICIFUGA

BUTTERCUP. See RANUNCULUS

BUTTERFLY WEED. See ASCLEPIAS

CACTUS. Tender perennials. Start seeds in sterile sand or artificial mix, and water sparingly. Some people like to cover the flat with a pane of glass or sheet of clear plastic, which is gradually removed as the seeds sprout. Different varieties start at different times, but most sprout slowly. Treat with fungicides, since damping-off diseases are likely to be a problem. A temperature of about 70 degrees F. should be maintained, and they should have full sun.

Cactus

Offshoots and cuttings root easily, but after they are cut from the parent they should be dried for several hours so they won't rot before rooting. Bottom heat is helpful, but high humidity is not needed.

Grafting is often used to produce combinations of different plants on the same cactus. Depending on the shape of the plant, they are either side-grafted or cleft-grafted and pinned in place until they grow together.

CALADIUM. Tender perennials often grown as indoor foliage plants because of their large, colorful leaves. The tubers can be split apart while they are dormant, just before they are to be planted. If cut

into pieces, each should contain at least two buds. They need warm temperatures to do well.

CALENDULA - POT MARIGOLD. Annual. Seeds germinate in six to ten days, at 70 degrees F. or warmer.

CALLA (*Zantedeschia*). Tender perennial. Most varieties are propagated by separating and planting the numerous offshoots that spring from the bulb-like rhizomes. The Golden Calla (*Z. elliottiana*) may be started from seed.

CALTHA - MARSH-MARIGOLD, COWSLIP. This wild flower is ideal for wet spots. The plants can be divided in early spring, or propagated by seed.

CAMPANULA - CANTERBURY BELLS, BELLFLOWERS. These hardy perennials and biennials are easily grown from seed, which should be planted early if the plants are to bloom the following year. They germinate in two to three weeks at 70 to 80 degrees F. The perennial varieties can be divided in early spring, and many kinds start well from cuttings.

CANDYTUFT. See IBERIS

CANNA. Tender perennial. Possible to start from seed, but the seeds are hard-coated and must be scarified, and the plants will not come true. Dividing the rhizomes in early spring is the usual method. They must be started inside for the earliest blooms.

CARDINAL FLOWER. See LOBELIA

CARNATION. See DIANTHUS

CASTOR BEAN. See RICINUS

CATNIP. See NEPETA

CELOSIA - COCKSCOMB. Annual. Seeds germinate in five-ten days at 70 to 80 degrees F.

CENTAUREA. Annuals and perennials, some tender, some hardy. Most kinds grow easily from seeds that germinate in two to three weeks at 70 to 80 degrees F. and perennial kinds can be divided easily in early spring. The Mountain Bluet (*C. montana*) spreads rapidly by underground runners and must be kept under control. Bachelor's Button or Cornflower (*C. cyanus*) is usually an annual, and seeds germinate in a week or so.

CERASTIUM TOMENTOSUM - SNOW-IN-SUMMER. Hardy perennial. Can be divided in the spring or fall, or started from cuttings taken in the summer directly after flowering. Plants layer easily, also. Seeds germinate in two or three weeks at 70 degrees F.

CERATOSTIGMA - PLUMBAGO. Divide the plants in early spring. Seeds and root cuttings are possible, also.

CHEIRANTHUS - WALLFLOWER. Half-hardy perennial-biennial. Can be potted to bloom in the house. Seeds germinate in two to three weeks, and must be planted inside very early to bloom the first year. Does not like an acid soil. Best plants are propagated by cuttings taken in early summer.

CHELONE - TURTLEHEAD. Starts easily from cuttings or the plant may be divided. Can also be grown from seeds.

CHINESE LANTERN. See PHYSALIS

CHRISTMAS CACTUS. See SCHLUMBERGERA

CHRISTMAS ROSE. See HELLEBORUS

CHRYSANTHEMUM. Annual and perennial. Most perennial varieties vary widely in hardiness, but many survive in the north with protection. Seeds start in two or three weeks at 70 degrees. Named varieties must be started asexually. Garden clumps should be divided each spring, and all start easily from cuttings taken anytime. These should be pinched back several times to get bushy plants.

CIMICIFUGA - BUGBANE. Perennial. Divide the plant in early spring, or plant seeds as soon as they ripen.

CINERARIA. See SENECIO

CLARKIA. Annual. Seeds germinate in one or two weeks over a wide range of temperatures. Extra light is beneficial for certain types, and they thrive where nights are cool.

CLEMATIS - PERENNIAL BUSH CLEMATIS. Divide the plants, or take cuttings in early summer. They like an alkaline soil and partial shade. (See Tree, Shrub, and Vine section for clematis vines.)

CLEOME - SPIDERFLOWER. Annual. Start inside for earliest blooms. Seeds germinate in a week at 70 degrees F. with additional light.

CLOCK-VINE. See THUNBERGIA

COCKSCOMB. See CELOSIA

CODIAEUM. See CROTON

COLCHICUM AUTUMNALE - AUTUMN CROCUS, SAFFRON. Hardy perennial. Separate corms to propagate. Seeds should be sown as soon as ripe, but may sprout slowly. Takes several years to bloom from seed.

COLEUS. Tender perennial. Plants with the best-colored foliage should be propagated by cuttings or by dividing the plants. Seeds germinate in about two weeks at 70-80 degrees F., and grow readily, but plants from seed show wide variation.

COLUMBINE. See AQUILEGIA

CONVALLARIA - LILY-OF-THE-VALLEY. Clumps can be broken up in either spring or fall and increased easily by dividing the pipes, as the buds at the ends of the rhizomes are called. Likes light shade and slightly acid soil.

CORAL BELLS. See HEUCHERA

COREOPSIS. Annual and perennial. Both grow readily from seed, which, if sown early, will bloom

the first year. Perennial clumps can be divided in spring or fall.

CORONILLA - CROWN VETCH. Perennial. Vigorous grower, often used for bank and ground covers. Can be grown from seed, which should be soaked in hot water for twelve hours and treated with an inoculant before planting, for best results. The plants can be easily layered and divided.

CORTADERIA SELLOANA - PAMPAS GRASS. Perennial. Propagated by dividing the clumps in early spring.

COSMOS. Annual. Seeds germinate in five to ten days at 70 to 85 degrees F. Extra light promotes fast growth. Most varieties bloom approximately two months from sowing.

CRANESBILL. See GERANIUM

CRASSULA ARGENTEA - JADE PLANT. Cuttings of this succulent start easily anytime from the rosettes at the tip of the branch. Dry for one day after cutting, before inserting in the medium.

CROCUS VERNUS - CROCUS. Hardy perennial with bulblike corms that increase naturally if leaves are allowed to die back each spring after blooming. Dig at that time, and separate. Can also be started from seed, but takes many years to bloom.

CROTON (*Codiaeum variegatum*). Tender perennial commonly grown as a housplant. Leaf cuttings start well in spring or early summer. Larger plants can be air-layered easily.

CROWN VETCH. See CORONILLA

CYCLAMEN. Tender perennial usually grown as a houseplant. Seeds are fine and need to be carefully spaced and barely covered. Seeds germinate in three to four weeks at 70 degrees F. Takes more than a year to flower from seed.

CYNOGLOSSOM - HOUND'S-TONGUE, CHINESE FORGET-ME-NOT. Biennial that is usually grown as an annual. It must be started inside early to bloom the first year. Seeds germinate in two to three weeks at 70 degrees F.

DAFFODIL. See NARCISSUS

DAHLIA. Tender perennials, sometimes treated as annuals. Small edging dahlias are grown from seed that should be started inside for early blooms. Germinates in one to two weeks at 70 to 80 degrees F. with extra light. Giant dahlias may also be grown from seeds, but named cultivars must be started asexually to come true. In the spring, divide tubers that have been stored over the winter, making sure there is a sprout on each piece. Cuttings of new growth root easily in early spring.

DAYLILY. See HEMEROCALLIS

DELPHINIUM. Hardy perennial. Seed should be planted as soon as it is ripe, or sealed in plastic and stored in a refrigerator until planting time. Seeds sown indoors in February will produce flowering plants the same year. Clumps of established plants can be divided in early spring, and cuttings taken from early summer growth may be divided. Seeds produce the huskiest plants, however. Hybrid varieties of delphinium are likely to be short-lived.

DIANTHUS. Hardy annuals, perennials, and biennials. There are many plants in this family including the pinks, carnations, sweet william, and sweet withersfield. Most start from seed with ease, and often self-sow. The perennial kinds can be divided in early spring, or started from cuttings. Layering is the easiest method of propagating *D. plumarius*. Most varieties of dianthus are short-lived.

DICENTRA-BLEEDING HEART. DUTCHMAN'S BRITCHES. Hardy perennial. Difficult to grow from seeds, and they need stratification for six weeks. Dividing the clump in early spring is an easier method of propagation. The old-fashioned variety, *D. spectabilis*, can be started from cuttings taken in early summer. Root cuttings taken after flowering can also be used. These should be three inches long or more.

DICTAMNUS-GASPLANT. Hardy perennial. The best way to propagate is to plant seeds as soon as they are ripe and before the seed coat hardens. If it has hardened, soak in hot water before planting. Division of the tough, carrot-like roots is difficult, and the plants do not enjoy being disturbed.

DIEFFENBACHIA. Tender perennial, often grown as a houseplant. New plants can be started by cutting a stem into sections two inches long when they are partially dormant, planting them horizontally, and then covering them with sand. Provide bottom heat. These will grow roots and tops. The plants can also be air-layered.

DIGITALIS-FOXGLOVE. Short-lived perennial, usually regarded as a biennial. Seeds planted as soon as they ripen in summer will bloom the following year. Those planted in the spring will not bloom until the second summer. Often the plant self-sows so freely that it becomes a weed in the flower border. Seeds germinate in ten to fifteen days.

DORONICUM-LEOPARDS-BANE. Hardy perennial. Plants may be divided in spring or in fall after they have finished flowering. Seeds germinate in two or three weeks, but inconsistently.

DRACAENA. Tender perennial often used as a pot plant. Older stems are cut off while they are dormant, the leaves removed, and the stem cut into small pieces and planted the same as dieffenbachia. Ends of shoots can also be air-layered or rooted as cuttings. Dip them in rooting powder, set them in moist sand, and root under lights. Sometimes offshoots at the base can be cut off and grown successfully.

ECHEVERIA. Tender succulent perennials used as houseplants in the North. These produce an abundance of rosettes that can be removed and planted in a sandy loam, where they root readily.

ECHINOPS-GLOBE THISTLE. Hardy perennial. Plants spread rapidly, and can be divided easily in the spring, in spite of their carrot-like roots. Root cuttings also grow well. Seeds, planted when ripe, grow easily, sprouting in ten to twenty days, but asexual propagation should be used for the best varieties, because they don't come true from seed.

ENGLISH DAISY. See BELLIS

EPIMEDIUM-BISHOPS-HAT. Hardy perennial. Divide the clumps in spring or fall.

EPIPHYLLUM - ORCHID CACTUS. These flowering cacti are tender perennials often grown as pot plants. Seeds grow well only after several months of storage. Can be propagated easily by cuttings placed in sand.

ERIGERON-FLEABANE. Annual and hardy perennial. Sow seeds in early spring for blooms the first year, or divide the clumps in spring or fall.

ERODIUM-HERONSBILL. Hardy perennial rock garden plant. Divide in early spring. Grows easily from seed, too.

ERYNGIUM-SEA-HOLLY. Hardy perennial. The plants seed themselves freely, and these can be dug and transplanted. Seeds can also be collected when ripe and planted at once to germinate the following spring. The plants can be divided, but not easily.

EUPATORIUM-Joe-Pye-Weed. Boneset. Hardy perennial. Native wildflower. Starts easily from either seeds or division.

EUPHORBIA-Spurge. Hardy perennial. Best propagated by division in spring or fall. See Trees, Shrubs, and Vines section for poinsettia.

EVENING PRIMROSE. See Oenothera

FERNS. For the amateur gardener, division is the easiest way to propagate ferns. Both the hardy outdoor and the potted ferns can be easily started by splitting up larger plants. Spring, before growth starts, is the best time to separate the rhizomes of houseplant ferns. Wild ferns can be dug in early spring and transplanted, if you have permission from the property owner. Plant them in a location very similar to the one where they were growing. The long list of ferns includes some that will grow nearly everywhere, but each variety is particular about sun, shade, moisture, and soil. Maidenhair, Christmas, and Fancy fern like the deciduous woods where they get an abundance of early spring sunshine, followed by heavy shade throughout the summer.

Adventuresome gardeners, or those who want to propagate a great many ferns, may gather and plant the spores. These are usually in spore cases on the underside of the fronds, and they should be checked with a magnifying glass to see if they are well developed and not empty. After gathering the fronds, dry them for a week in a warm place. Then place them on a sterile medium (two-thirds peat, and one-third perlite or vermiculite is good), cover them only with a piece of clear plastic or glass, and place under light. Keep the medium moist at all times, but do not allow too much moisture to condense on the glass, or mold will form. Since part of the spores are male and part female, fertilization must take

Fern

place on the medium, so it will take from a few days to a few weeks before they germinate and grow.

The new plants should be carefully transplanted while they are quite small, and a pair of tweezers is almost a necessity. As they grow, a second transplanting will be necessary to space out the plants so they will have enough room. In two years the plants should be large enough to use as houseplants.

A few ferns, such as the bladder fern *(Cystopteris bulbifera)*, produce bulbils on the underside of their fronds which can be gathered and planted. These also take several weeks to grow, and about two years before they develop into a sizeable plant.

FICUS ELASTICA-Rubber Plant. Tender perennial. Cuttings from young plants can be rooted in the spring in a warm location. Sand or perlite, kept moist, is a good medium. Larger plants may be propagated by air layering, a method that is often used.

FLAX. See Linum

FLEABANE. See Erigeron

FLOWERING TOBACCO. See Nicotiana

FORGET-ME-NOT. See Myosotis

FOUR O'CLOCK. See Mirabilis

FOXGLOVE. See Digitalis

FREESIA. Tender perennial used as a greenhouse plant in the North. It can be dug in the fall and the small cormels removed and planted in a warm greenhouse. Larger cormels can be potted at once and smaller ones placed in flats of soil to grow for potting in the spring. Offsets may also be taken from the plant during the winter, and potted. Seeds can be planted in spring, and the young seedlings wintered over in a warm greenhouse for spring potting.

GAILLARDIA-Blanketflower. Hardy perennials, biennials, and annuals. Perennial kinds can be divided in the spring, but the plants are short-lived. Both annuals and perennials start easily from seeds which germinate in ten to fifteen days at 70 degrees F. Perennial kinds bloom the second year after planting. Root cuttings are also used.

GALANTHUS-Snowdrop. Hardy perennial. Bulbs can be dug in the fall and divided, or offsets can be separated from the main plant and planted.

GASPLANT. See Dictamnus

GENTIANA-Gentian. Most of these are hardy perennials that bloom late in the season, and in the North may not always ripen their seed before the first frost. Seed should be planted as soon as ripe, remain in the ground over the winter, and sprout in the spring. The fringed gentian is a biennial and rather difficult to grow. The bottled gentian likes an acid soil, so keep lime and alkaline water away from the plants. Certain varieties, *G. andrewsii* and *G. seponaria*, can be divided in the spring.

GERANIUM. See Pelargonium

GERANIUM-Cranesbill. Should not be confused with the pelargoniums, which are houseplant geraniums. The cranesbill is a hardy perennial which can be easily started from seed. Germination takes place in a few weeks. The plants can also be divided in the spring, and the better varieties should be started in that way. Cuttings may be made in the summer.

GERBERA JAMESONI TRANSVAAL DAISY. Tender perennial that is often grown in greenhouses in the North. Can be started from seeds that germinate in ten to fourteen days at 70 degrees F. Sow seeds indoors in January, for setting out after frosts are over. Basal offshoots may be rooted as cuttings in early spring.

GEUM. Hardy perennial. Divide clumps in spring or fall. Seeds germinate in three to four weeks at 70 to 85 degrees F.

GINSENG. Hardy perennial. Plants are difficult to divide or start from cuttings. Seeds should be planted in the fall as soon as ripe, or stratified and planted in the spring. Woods soil or compost is the best medium when planting in flats or pots, but the seed can be planted directly in the woods if you wish. Mice sometimes eat the newly planted seed, so it's best to start them in flats and transplant them outside later. Both seedlings and growing plants need a heavy shade provided either artificially or by trees, and they tend to grow slowly.

GLADIOLUS. Tender perennial. Seeds are used only to develop new varieties. Corms are dug in the fall, dried, and stored for the winter. In the spring they are separated and planted outside. The smallest corms will not bloom the first year.

GLOBE FLOWER. See TROLLIUS

GLOBE THISTLE. See ECHINOPS

GLORIOSA. See RUDBECKIA

GLOXINIA. See SINNINGIA

GODETIA-SATIN-FLOWER. Annual. They need over a hundred days before blooming begins, so should be started indoors. They germinate in ten to twenty days at 70 to 85 degrees F.

GOUTWEED. See AEGOPODIUM

GRAPE HYACINTH. See MUSCARI

GRASSES, ORNAMENTAL. Grasses with variegated foliage do not come true from seed, so are best propagated by divisions in early spring. Seed is the best way to propagate the other grasses, and it is important that the seedlings be transplanted early, to give them plenty of room to grow. Grass plants that are overcrowded do poorly.

GYPSOPHILA. Annual and perennial. Seed germinate in ten to twenty days at 70 degrees F., and the plants can be divided. Certain varieties with double or colored flowers, such as Bristol Fairy, are cleft-grafted on roots cut from seedlings. The grafts must then be put in a shaded cold frame or propagation case until they heal. Perennial gypsophila can also be started by cuttings taken in early summer, treated with a rooting chemical, and put under mist.

HEATH and HEATHER. See Trees, Shrubs, and Vines section.

HELENIUM-SNEEZEWEED. Hardy perennial. Can be divided in early spring, or started from seed that germinates in one or two weeks.

HELIANTHEMUM - SUNROSE. Shrubby, half-hardy perennial. Softwood cuttings taken in the spring root well. Grows easily from seed which germinates in ten to twenty days at 70 to 80 degrees F. Divisions of the plant are usually short-lived.

HELIANTHUS-SUNFLOWER. Annuals and perennials. Annual varieties are started by seeds that germinate in two to three weeks at 70 to 80 degrees F. Perennial kinds are divided in early spring, or started from cuttings taken in early summer.

HELIOPSIS-ORANGE SUNFLOWER. Hardy perennial. Clumps can be divided in fall or spring. Seeds germinate in about ten days at 70 degrees F., but seedlings do not come true.

HELIOTROPIUM-HELIOTROPE. This is a tender perennial that is usually grown as an annual. Seeds should be started indoors for longest season of bloom. Germinate in two to four weeks at 70 to 85 degrees F. Also propagated by cuttings rooted under moist cool conditions, and by layers in the spring.

HELLEBORUS-CHRISTMAS ROSE. Hardy perennial. Clumps may be divided in early spring or fall, but because the roots are brittle, handle them carefully. Seeds should be sown as soon as ripe, and will germinate the following spring, but probably not blossom for a least two years.

HEMEROCALLIS-DAYLILY. Hardy perennial. Seeds are used only for developing new varieties and should be planted as soon as ripe. Clumps can be split anytime, but this is best done in spring or fall. Each division should have at least three sprouts.

HENS and CHICKENS. See SEMPERVIVUM

HERBS. The variety of herb plants is large. It includes annuals and tender and hardy perennials. Most can be grown easily from seed, and the perennial kinds usually start well from cuttings, or the plants can be divided or layered. Some perennials are extremely short-lived, while others, such as the mints, spread more rapidly like weeds and live many years.

The following annual varieties are usually grown from seed: anise, basil, borage, caraway (biennial), chamomile, chervil, coriander, dill, fennel, marjoram, parsley (biennial), peppergrass, and summer savory.

Because most perennial herbs start easily from divisions, cuttings, or layering, they are usually propagated by these methods. However, with the exception of

French tarragon, which must be propagated asexually, it is possible to start most of them from seed, including members of the allium family (chives, garlic, leeks, and shallots, though these are usually propagated by division of their bulbs), angelica, burnet, catnip, costmary, curley mint, hops, horehound, hyssop, lavender, lemon balm, lovage, oregano, peppermint, rosemary, rue, sage, winter savory, sorrel, spearmint, sweet woodruff, Russian tarragon, thyme, watercress, and wormwood.

HERONSBILL. See ERODIUM

HESPERIS-SWEET ROCKET. This biennial flower has escaped from gardens in many areas and become almost wild because it self-sows so readily. Ordinary plants can be easily propagated by seeds, but the best varieties should be started from divisions or cuttings so they will come true.

HEUCHERA SANGUINEA-CORAL BELLS. Hardy perennial. Can be divided in early spring, or started from seed, which germinates in two or three weeks. Seed should be started inside in early spring, because the seedlings grow slowly. Leaf cuttings, taken in mid-summer, with part of the stem, root easily.

HIBISCUS-MALLOW. Annual and moderately hardy perennial. Can be grown from seed planted inside in early spring. For propagating special varieties, take hardwood or softwood cuttings. The roots can also be divided.

HIPPEASTRUM-AMARYLLIS. Tender perennial, grown as a houseplant. It is possible to propagate them from seeds, but they germinate slowly and unevenly, and take many years to bloom. The offsets may be removed and planted, and these should bloom in about two years.

HOLLYHOCK. See ALTHAEA

HORSERADISH (*Armoracia rusticana*). Hardy perennial herb. Roots can be cut apart in early spring or late fall. You can also cut the tops off roots that have been dug for eating or storage, and plant them, with a bit of root attached. These will quickly grow into new plants. Root cuttings are also possible.

HOSTA - PLANTAIN LILY, FUNKIA. Hardy perennial. These plants can be divided anytime, but most easily in spring or fall. They like cool shady places with slightly moist soil.

HUNNEMANNIA - GOLDEN-CUP. Tender perennial that is usually treated as an annual. Needs to be started inside early to bloom the first year. Seeds germinate in fifteen to twenty days.

HYACINTHUS - HYACINTH. Hardy perennial grown from a bulb. New bulb offsets can be separated, and planted in the fall. Commercially the bulbs are scooped or notched to stimulate the growth of bulblets. See Chapter III. Seeds are used to create new varieties, but it takes many years from seed to bloom.

HYPERICUM - ST. JOHNSWORT, AARONSBEARD. Divide in early spring, or start from cuttings in early summer. May be grown from seed planted anytime. (For shrub hypericum, see Trees, and Shrubs, and Vines section.)

IBERIS - CANDYTUFT. Annual and hardy perennial. The best perennial plants are propagated by divisions and cuttings. They can be divided in early spring, or started from cuttings taken in early summer. Both annual and perennial varieties start easily from seeds which germinate in a week at 70 degrees F. Perennial seedlings are likely to be weak-looking for the first year or two.

IMPATIENS - BALSAM. Annuals and partially hardy perennials. The seeds should be started inside in late

winter for a long period of bloom. They usually germinate within a week at 70 degrees F. when given additional light. Easily started from cuttings in the summer.

INDIGO. See BAPTISIA

IPOMOEA - MORNING GLORY. Tender perennial and annual. Seeds should be soaked overnight or longer in warm water before planting, or their hard coats should be nicked with a file. Germinates in one to three weeks at 70 degrees F. When starting plants indoors, planting in peat pots prevents transplanting problems.

IRESINE - BLOODLEAF. Tender perennials used as houseplants and bedding plants. They may be wintered over in a greenhouse and used as stock plants for cuttings in late winter.

IRIS. Perennial. There are many classes of iris, and most can be easily propagated by division of their rhizomes or bulbs. The bearded iris, which is commonly found in gardens, is propagated by dividing the rhizomes after the blooming season is over. Use only the side shoots. Japanese iris is propagated by splitting up the clumps in early spring. Bulbous types are dug in the dormant season, the old bulb discarded, and the small offshoots separated and planted out. Seeds are difficult to germinate and should be frozen. They are used primarily to develop new varieties.

JACOB'S LADDER. See POLEMONIUM

JADE PLANT. See CRASSULA

JOE-PYE-WEED. See EUPATORIUM

KALANCHOE. These are succulents, often used as houseplants. They can be started by seeds, but cuttings produce better plants. Regular stem cuttings, offshoots, or leaf cuttings are all possible. They should be left out to dry a few hours after detaching from the plant so their sticky sap won't make them rot. The bulbils that develop along the leaves of certain kalanchoes can also be used for propagation. Most mature plants can be dug up and separated into several plants.

KNIPHOFIA - TRITOMA, TORCH-LILY. Partly hardy perennials that need winter protection in the North. Seeds germinate in twenty to twenty-five days at 70 to 85 degrees F. and grow slowly. Large clumps can be lifted and divided in the spring.

KOCHIA - SUMMER CYPRESS. Annual. Seeds start so easily that it may become a weed. Germinates in five to ten days at 70 to 80 degrees F., and should be started inside to get the quickest effect for summer.

LACEFLOWER. See TRACHYMENE

LANTANA. Tender perennial. Seeds germinate in six to seven weeks. Cuttings root rapidly in early summer, and should be used to propagate named varieties.

LATHYRUS - PERENNIAL PEA VINE, SWEET PEA. Hardy perennial and annual. *L. latifolius* plants may be divided in early spring. Seeds germinate in fifteen to twenty days.

Annual sweet pea (*L. odoratus*) should be planted as early as possible in the spring after the ground thaws. Soak the seed in warm water for several hours before planting to speed up germination. May also be started inside in peat pots and transplanted for early blooms.

LAVANDULA - LAVENDER. Moderately hardy perennial. Cuttings can be taken in late summer or early fall, then rooted and wintered in a greenhouse. Plants may also be divided in early spring, or layered. Seeds germinate in about two weeks.

LAVATERA. This annual member of the mallow family is propagated by seeds sown early inside, or outside after all danger of frost is past.

LIATRIS - GAYFEATHER. Hardy perennial. Divide in early spring. Seeds germinate in three to four weeks. Likes poor soil that is a bit moist.

LILIUM - LILY. Hardy and half-hardy perennials, grown from bulbs. Various kinds of division are the most common methods of propagation. For information about dividing the bulbs, separating bulblets and bulbils, and scaling, see Chapter III (Division).

Seeds usually result in hybrid plants and are used for originating new varieties and those species, such as the Regal, that come true from seed. Seeds of some varieties, such as the Mid-Century hybrids, germinate within six weeks, but others, like the Canada lily (*L. canadense*) take up to a year and a half to germinate. Dust the seed with a fungicide, and plant it indoors in the winter in deep flats, so they do not need to be moved for at least a full year.

LILY-OF-THE-VALLEY. See CONVALLARIA

LIMONIUM - STATICE, SEA-LAVENDER. Annual, biennial, and perennial. Sow indoors and transplant after frost danger is past. Perennial sea-lavender (*L. latifolium*) may also be divided.

LINARIA - TOADFLAX. Annuals and perennials. Perennial clumps spread by underground stems. May be divided in fall or spring. Seeds germinate easily in three to four weeks and often self-sow.

LINUM - FLAX. Annual and hardy perennial. Perennials may be divided. Both may be grown from seed, which germinates in three to four weeks.

LOBELIA - CARDINAL FLOWER and others. Annual and hardy perennial. Divide perennial clumps in early spring. Cuttings may be taken from plants that were potted in fall and wintered over in a greenhouse. Seeds germinate in twenty-five to thirty days at 70-80 degrees F. Start annual seed indoors early.

LOBULARIA - SWEET ALYSSUM. Perennial in many places, but usually grown as an annual from seed. Germinates in five days.

LUNARIA - HONESTY, MONEYPLANT. A biennial grown principally for its silvery seed pods, which are used in dried arrangements. Grows easily from seed, which germinates in two to three weeks. Should be planted where it is to grow, in partial shade if possible, or in peat pots since it doesn't transplant easily. Sometimes self-sows.

Lupine

LUPINUS - LUPINE. Annuals, hardy perennials. Annuals are propagated by seed, which germinates in two to three weeks. Perennial clumps may be divided in early spring. Some of the best varieties can also be grown from seed, which germinates in three to four weeks. Sometimes escapes and becomes a wildflower.

LYCHNIS - CAMPION, MALTESE CROSS. Hardy perennial. Clumps may be split up in fall or spring. Seeds germinate in two to three weeks.

LYCORIS - MAGIC LILY, NAKED LADY. Periennal. Small bulblets are separated from the main bulb when they are dormant in late fall, after blooming.

LYTHRUM - LOOSESTRIFE. Perennial. Easy to propagate by division, stem cuttings, or seed.

MALLOW. See HIBISCUS

MARIGOLD *(Tagetes).* Tender annual, biennial, or perennial usually grown from seed. Start seeds inside for early summer bloom. Germinates in a few days at 70-80 degrees F.

MARSH-MARIGOLD. See CALTHA

MATTHIOLA INCANA - STOCK. Annual, biennial, or perennial, usually grown as an annual. Start seeds inside for early summer bloom. Germinates in about ten days.

MERTENSIA - VIRGINIA BLUEBELLS. Established clumps can be divided in fall or early spring. Seeds planted in the spring should bloom the following year.

MICHAELMAS DAISY. See ASTER

MIRABILIS - FOUR O'CLOCKS. Tender perennials usually treated as annuals. Easily grown from seed, which germinates in one to two weeks. Sow outdoors, when danger of frost is past, or indoors for early bloom. Plants vary greatly, and some produce several colors on the same plant. Best ones can be treated like dahlias, that is, dug in early fall, dried, and stored in a cool place, divided and set out again in the spring, where they will bloom early.

MONARCH DAISY. See VENIDIUM

MONARDA - BEE BALM. BERGAMOT. Hardy perennial. Plants spread rapidly, mint-fashion, and can be separated easily at nearly any time. Also grown readily from seed.

MONKSHOOD. See ACONITUM

MORNING GLORY. See IPOMOEA

MOSS ROSE. See PORTULACA

MUSCARI - GRAPE-HYACINTH. Hardy perennial that grows from small bulbs. Spreads rapidly and is readily propagated by dividing the offshoots in the fall.

MUSHROOMS *(Agaricus).* The growing of mushrooms from spores is not easy, and special environmental conditions must be met, including proper humidity, temperature, and ventilation. The spores develop a mass called spawn, which develops into edible mushrooms. Most growers purchase spawn from specialized laboratories. It must be cultured in an organic medium, which was once a compost made entirely from horse manure, but is now a combination of horse manure and a synthetically produced compost. The compost is pasteurized in an exacting process for ten days, and the spawn added. Growing is usually done in cellars, caves, or special concrete buildings that have been slightly pressurized so that no unwanted plant spores can easily enter.

MYOSOTIS - FORGET-ME-NOT Hardy biennial. Plants often self-sow, and can become very weedy. Seed germinates in two to three weeks and is easily started outdoors.

MYRTLE. See VINCA

NARCISSUS - DAFFODIL, JONQUIL. Hardy perennial grown from bulbs. Seeds are used only for propagating wild types, and for originating new varieties. Bulb clumps are divided after blooming. (See Bulbs in Chapter III, Division.)

NASTURTIUM *(Tropaeolum).* Perennial and annual. Seeds may be started inside in peat pots for earliest blooms; otherwise plant them where they are to grow, as they transplant with difficulty. Seeds germinate in about two weeks. Exceptionally good kinds are sometimes propagated by cuttings.

NEPETA - CATNIP. Hardy perennial. Certain varieties produce flowers of an outstanding blue color. The plants, which spread rapidly, may be divided in early spring. Seeds and cuttings start readily, too. It can also be layered, to produce even more plants.

NICOTIANA - FLOWERING TOBACCO. Annual, biennial, or perennial. Seeds should be started inside for earliest blooms. Germinates in about a week at 70-80 degrees F., with additional light.

NIEREMBERGIA- CUPFLOWER. Tender perennial, usually treated as an annual in the North. Sow seeds in February, and plant outside after frost danger is over. Seeds germinate in fifteen to twenty days. Use asexual propagation to retain good color and form of new plants. Divide clumps in the spring. Cuttings taken in early summer root easily, too.

NYMPHAEA - HARDY WATERLILY. Nearly all can be started by separating offshoots from the tubers of established plants in the spring. Replant only young, vigorous-looking tubers by planting them horizontally where you want them to grow, making sure each has at least one good bud; or plant them in tubs or buckets of soil which are then sunk into a shallow pond or pool. A layer of sand over the soil in the bucket helps keep the water in the pool clean.

The seeds of wild waterlilies may be planted in flats of sandy soil, and immersed in shallow water.

OENOTHERA-EVENING PRIMROSE. Hardy perennial or biennial. Divide the clumps anytime, but it is best done in early spring. Also grows well from seeds that germinate in ten to fifteen days.

ORCHIDS. The orchid family is a large one, and there are wide variations in the six major groups. They have long been the caviar of the floral industry, and even during Depression times the flowers sold at unbelievably high prices. Asexual propagation is usually slow and difficult, although offshoots of the bulbs may be used with some species, and air layering or cuttings with others. Plants grown from cuttings and other asexual methods are weak, produce few blooms, and are short-lived.

Seeds are used for growing many plants commercially and for developing new varieties. A seed capsule takes up to a year to mature before the seed ripens, and the seedling plant may take five or more years to bloom. The discovery of tissue culture opened a new dimension in orchid growing, making the plants easy to reproduce. With tissue culture, rare, choice, and new varieties can be quickly reproduced by the thousands in the laboratory.

Although the culture of orchids is not difficult as long as a high temperature and humidity are maintained, propagation is most difficult for the home gardener, and is best left to the specialists.

PAEONIA-PEONY. Division is the best way to increase these valuable perennials. This is ordinarily done in late summer or early fall, after blooming is over. It can also be done in spring, though the plants may not bloom the same summer. Peonies being grown for commercial production may be divided every two years if they are growing well, but home gardeners will probably not want to separate them any more often than once in five years, because it disturbs the plant so much.

Dig the whole root carefully and cut it apart, making sure there are at least three eyes, or sprouts, on each root. Plant each section with an eye only a half-inch below the surface of the soil. When planted too deep, they will not bloom for many years.

Some gardeners do not like to disturb their plants, so, instead of digging the whole plant, they leave part of it in the ground untouched. To propagate in this way, chop off the outside portions with a sharp spade, making sure there are

sprouts on each piece. Leave some soil on the roots, if possible, to reduce the transplant shock.

Large pieces of eyeless peony roots, three inches or longer, can be planted vertically, right side up, about two inches below the surface of the soil. The best time to do this is in early spring. In time, many of them will grow.

Peonies are heavy feeders, and both the young and old plants respond to an abundance of manure piled over them in the fall. This is especially important if you want to produce a great many new plants.

The propagation of tree peonies is covered in the section on Trees, Shrubs, and Vines.

PAINTED DAISY. See PYRETHRUM

PAMPAS. See CORTADERIA

PANSY (*Viola tricolor hortensis*). Hardy, but short-lived perennial, often treated as an annual or biennial. Although they are usually grown from seed, the best varieties are propagated by division or cuttings, since the seed doesn't always come true. Seed germinates in ten to fifteen days, and is often planted inside in mid-winter to produce plants that can be set out in early spring. It may also be sown in the ground in summer, preferably in raised beds, for plants that will flower the following year. A mulch helps them over-winter.

Blossoms should always be kept picked unless seed is being grown. Plants that produce seed are short-lived, and are treated as biennials. If you save your own seed, make sure there are no wild pansies nearby, or the seedlings will produce mostly small, violet-type blooms.

PAPAVER-POPPY. Annuals, hardy perennials. All poppies are easily raised from seed, but the best colors in the perennial varieties don't come true. Seed usually germinates in a week or two at 60 degrees F. and is best planted where the plants

Oriental poppy

are to grow or in peat pots, since poppies do not transplant well. Seed of the Oriental poppy (*P. orientale*) should be barely covered and sown thinly. Iceland poppies (*P. nudicaule*) are usually treated as a biennial, and self-sow generously.

Oriental poppies can be propagated asexually by dividing the plants directly after they finish blooming and die down. Root cuttings taken in the fall, cut in three-inch lengths and planted horizontally about an inch deep in sandy soil, make good plants by late spring. Plant them in a cold frame or flat and mulch them over the winter, for best survival. Root cuttings may also be made in the spring. It is best to plant them in medium-size pots for easy transplanting during the summer.

PELARGONIUM-GERANIUM. Tender perennial that is usually grown as a summer outdoor bedding plant or indoors as a pot plant, although in frost-free areas they thrive outside year-round. Stem cuttings three or four inches long root easily. Remove the foliage on the bottom of the stem. Three or four small leaves toward the top are all that are needed. Because of the high sap content in the stems, the cuttings are less likely to rot while rooting if they are left to dry, out of the sun for an hour or more before placing them in a sterile medium. We have had the best results using a perlite-vermiculite mix. Rooting in small peat pots prevents damage that might be caused by separating plants rooted in flats. Water them thoroughly initially, and then keep them rather dry, and provide a fairly warm temperature. Pot them in soil as soon as roots form. Of course, it is also possible to root a small number of geraniums on a windowsill by placing a "slip" in a jar of water.

Many varieties start well from seed. The colors are quite uniform, but germination is often irregular. They usually bloom later than plants started from cuttings. Germination takes from two to four weeks at 75 to 80 degrees F.

PENTSTEMON-BEARDTONGUE. Half-hardy perennial that is sometimes treated as an annual. Clumps can be divided in spring or fall, and side shoots taken as cuttings in mid-summer root fairly readily. Seeds germinate in ten to fifteen days at 75 degrees F.

PEONY. See PAEONIA

PEPEROMIA. Tender succulent perennial often used as a houseplant. Divide, or start from stem cuttings or leaf cuttings.

PERIWINKLE. See VINCA

PETUNIA. Tender perennial, usually grown as an annual. Seeds should be started inside early both for a long season of bloom and because the small seeds are difficult to plant successfully outdoors. Do not cover the seed when planting, but press it gently into the medium. They germinate in a week or two at 70 degrees F., and benefit from extra light. Soft cuttings root rapidly when taken nearly anytime.

PHILODENDRON. Tender perennial vine usually grown in the North as a houseplant. Stem cuttings root easily. Take portions with two or more joints and place them in sand or soil in a warm

place, or root them in a glass of water. They can be layered, and large plants can be air layered. Seeds germinate fairly quickly if planted as soon as they mature, before they dry out.

PHLOX. Annual and hardy perennial. Seeds germinate in ten to fifteen days at 70 degrees F., and annual varieties (P. drummondi) should be started inside for early blooms.

Seeds of perennial varieties (P. paniculata and others) do not come true. They revert mostly to magenta colors, and only rarely produce worthwhile plants. Since the plants self-sow so easily, flowers should be picked as soon as they fade, or inferior seedlings will sprout, which will quickly crowd out the good plants.

The plants can be divided either in spring or fall, and cuttings taken in early summer root easily. For large numbers of plants, root cuttings can be used. Dig the entire plant in the fall, and cut off all the large roots except those close to the crown. The crown can then be replanted. Cut the roots into pieces about two inches long, and plant them in flats of rich, sandy soil, a half-inch deep and mulch heavily or store in a protected cold frame for the winter. By late June they should be ready to transplant.

If you want only a few plants, an easier way is to leave the original clump in place. Cut all around it, severing the outside roots with a sharp spade, but leave the roots in the ground. This, too, is best done in the fall. The following spring, the roots that were cut off send up new little plants which can be dug and transplanted in early summer.

The ground, or creeping, phlox (P. sublata) may be started by division, layering, or from cuttings.

PHYSALIS-CHINESE LANTERN PLANT. Hardy perennial. Plants often self-seed, and may become weedy. Small seedlings can be

transplanted, or seeds collected and planted as soon as they are ripe. Clumps of older plants can be divided in spring or fall, and root cuttings grow easily.

PHYSOSTEGIA-FALSE DRAGONHEAD. Hardy perennial. Divide clumps in early spring. Seeds germinate easily also.

PIGGYBACK PLANT. See TOLMIEA

PILEA - ALUMINUM PLANT, ARTILLERY PLANT. These houseplants are easily propagated by cuttings.

PLANTAIN LILY. See HOSTA

PLATYCODON-BALLOON FLOWER. Hardy perennial. The fleshy root can be divided, but with difficulty, and should be attempted only in the spring. Seeds germinate in ten to twenty days. Cuttings treated with a rooting chemical start easily under mist in early summer.

PLUMBAGO. See CERATOSTIGMA

POLEMONIUM-JACOB'S LADDER. Hardy perennial. Plants can be divided in early spring or late summer. Seed germinates best if planted as soon as it is ripe in early fall, but if this is not possible, spring-planted seed also does fairly well.

POLIANTHES-TUBEROSE. Tender perennial grown from a bulb. Dig and divide offsets after blooming.

POPPY. See PAPAVER

PORTULACA-MOSS ROSE. Annual which sometimes reseeds itself. Seeds should be started inside for earliest blooms. Germinates in ten to fifteen days and grows faster with additional artificial light.

PRIMROSE. See PRIMULA

PRIMULA-PRIMROSE. Hardy perennial. To propagate, divide the plants in the fall or in the spring, following their blooming. Usually they are grown from seed, which is very fine and needs careful handling. Some seeds germinate within a few weeks, but others may take several months. It is best to sow the seed as soon as it is ripe in late summer, but if that is not possible, place it in the freezer and then let it thaw and freeze repeatedly for a few days just before planting. Primrose plants do best in slightly acid soil well supplied with humus.

PYRETHRUM (*Chrysanthemum coccineum*) PAINTED DAISY. Annual, perennial. Propagate by dividing the plants in spring or late summer, or by seeds sown in the spring.

RANUNCULUS-BUTTERCUP. Hardy perennial. Plant seed in the spring or as soon as ripe in summer. *R. asiaticus* grows from tuberous roots, which can be divided in spring or fall.

RHUBARB (*Rheum rhabarbarum* or *R. rhaponticum*). Hardy, herbaceous food plant. Grows easily from seed, which can be planted as soon as ripe, or saved and planted in spring. Best varieties must be started by asexual means. Dividing the clump in spring or late fall is the usual method, and if they are to stay productive, the plants should be divided every five to seven years. The whole root can be lifted and divided, or several outside sections can be split off with a sharp spade and removed, leaving the interior of the plant intact. Each section should include one or more eyes. The new divisions should be planted two to three inches deep. They respond favorably to large amounts of manure.

RICINUS COMMUNIS-CASTOR BEAN. Annual. Germination of the seed can be speeded up by soaking the seed in barely warm water for twenty-four hours or by nicking the hard seed coat with a file. Start plants inside in peat pots, and set them out after frost danger is past, to get the largest plants possible before fall frost.

RUBBER PLANT. See FICUS

RUDBECKIA-CONEFLOWER, BLACK-EYED SUSAN, GLORIOSA DAISY, GOLDEN GLOW. Annual, perennial, or biennial. Gloriosa daisies are the most spectacular of the group, and are easily grown from seed. Seed germinates in one to three weeks and if started inside in early spring will bloom the first year. Perennial rudbeckia may be divided, and some varieties self-sow.

SAINTPAULIA-AFRICAN VIOLET. Tender perennial, usually grown as a pot plant. Best varieties should be propagated by division or by leaf cuttings that are taken from the middle of the plant. They will root in water or other rooting medium. Seeds are very small and need special care since they are susceptible to damping-off diseases and germinate only at high temperatures. Plants grown from seed will not be like their parent.

SALPIGLOSSIS SINUATA-PAINTED TONGUE. Annual. Seeds are small and should be started inside. Even under the best of conditions, they germinate irregularly. Seedlings should be transplanted into peat pots for easier planting outside later.

SALVIA-SAGE. Annual, biennial, perennial. Seeds are best started inside for early blooms. Germinates in one to two weeks at 70 degrees F. The hardy perennials may be propagated by divisions. Softwood cuttings work well, but seed ordinarily produces the best plants.

SANSEVIERIA-SNAKEPLANT. Tender perennial, often grown indoors in pots and tubs. The rhizomes of all varieties except *S. laurentii* can be easily divided. The leaves, cut into sections three or more inches long, root easily in a perlite-vermiculite mixture. These grow roots and a new shoot at the base of each cutting.

SAPONARIA-BOUNCING BET. Annuals and hardy perennials. Can be divided easily, as the plant spreads with much vigor. Also easy to propagate by seeds and cuttings.

SATIN-FLOWER. See GODETIA

SAXIFRAGA-This family encompasses a large group of rock garden plants which are annual, biennial, and perennial. Many can be grown from seed, which germinates easily. Named varieties should be started by asexual methods. They may be divided, and cuttings of the small rosettes, taken after they finish blooming, root well. The strawberry geranium (*S. stolonifera*) can be propagated easily by rooting its runners.

SCABIOSA-PINCUSHION FLOWER. Annual, biennial, and perennial. Annual varieties can be started easily by seeds. Perennial kinds are started either by seeds, which come quite true, or by divisions.

SCHEFFLERA (*Brassia actinophylla*). Shrub, used as a houseplant. Propagated by air layering.

SCHIZANTHUS-BUTTERFLY FLOWER. Annual. Seeds should be started inside in February for early blooms outdoors in summer, but can also be planted in fall for indoor flowering in late winter. Germinates in ten to fifteen days at a fairly cool temperature.

SCHLUMBERGERA-CHRISTMAS CACTUS. Houseplant that roots easily from cuttings. Sometimes it is grafted on another cactus as a novelty.

SCILLA-Hardy perennials that are grown from bulbs. Dig after blooming and separate the small bulblets from the parent plant.

SEA-HOLLY. See ERYNGIUM

SEDUM-STONECROP. Low-growing succulents that are easily propagated by divisions, and by cuttings which should be allowed to dry out for a day or more before planting in sand. Can become very weedy, since they reproduce naturally so rapidly.

Hens and chickens

SEMPERVIVUM-Hens and Chickens. Some of these are hardy perennials, while others are tender and grown mostly as pot plants in the North. The small rosettes can be divided, and root easily. They may also be grown from seed.

SENECIO-Cineraria, Groundsel. Annuals and tender perennials. Annuals are usually grown from seed, and cinerarias are often used as pot plants for winter blooms. Seeds germinate easily in two to three weeks. Perennial varieties are grown from seed, division, and cuttings.

SINNINGIA-Gloxinia. Tender perennial, usually grown as a pot plant. The tuberous roots can be divided to start more plants. Leaf or stem cuttings are used when large numbers are wanted. Seeds do not come true, but are useful for growing a variety of colors and types. The fine seed requires careful planting and care in watering and transplanting the tiny seedlings. Plant them sparingly and barely cover the seed. Should germinate in fifteen to twenty days at 70 degrees F.

SNAKE PLANT. See Sansevieria

SNAPDRAGON. See Antirrhinum

SPIRAEA. See Astilbe

SPURGE. See Euphorbia

STATICE. See Limonium

STOCK. See Matthiola

STOKESIA-Stokes Aster. Hardy perennial. Clumps can be divided in early spring, and also started from root cuttings. Seeds germinate in four to five weeks.

STRELITZIA REGINAE-Bird-of-Paradise. Tender perennial. Small offsets can be removed and planted, or the rhizomes can be divided in spring. Seed should be planted soon after ripening, before the seed coat hardens. Germinates only at quite warm temperatures.

SUNFLOWER. See Helianthus

SUNRAY. See Acroclinium

SUN-ROSE. See Helianthemum

SWEET ALYSSUM. See Lobularia

SWEET PEA. See Lathyrus

SWEET WILLIAM. See Dianthus

TAGETES. See Marigold

THALICTRUM-Meadow-Rue. Hardy perennial. Plant can be divided in spring or fall. Seeds germinate in several weeks.

THUNBERGIA-Clock-Vine. Tender perennial usually treated as annual. Seedlings grow slowly, and should be started inside for best results. Cuttings root easily and grow well.

THYMUS-Thyme. Hardy perennial. Most plants spread rapidly and are easily started from divisions or layers. Seed also grows well.

TITHONIA ROTUNDIFOLIA-Mexican Sunflower. Tender perennial, usually grown as an annual. Seeds germinate in ten to twenty days and should be started indoors for setting out after frost danger is over.

TOLMIEA MENZIESI Piggyback Plant. Easy to propagate. The new plantlets form at the base of the leaves. These can be picked off and planted any time.

TORCHLILY. See Kniphofia

TRITOMA. See Kniphofia

TRACHYMENE CAERULEA-Blue Laceflower. Tender perennial, often treated as an annual. Sow the seeds early inside for blooms early in the summer.

TRADESCANTIA FLUMINENSIS-Wandering Jew. Houseplant easily propagated by divisions or cuttings taken at any time.

TROLLIUS-Globeflower. Hardy perennial. Divide in early spring. Seeds may be planted in fall, but named varieties will not come true.

TROPAEOLUM. See Nasturtium

TUBEROSE. See Polianthes

TULIPA-Tulip. Hardy perennial grown from a bulb. Dig bulbs after blooming and separate offsets. Seeds are used only to develop new varieties. (See Bulbs in Chapter III, Division.)

TURTLEHEAD. See Chelone

VALERIANA-Valerian. Hardy perennial, sometimes used as a medicinal herb. Divide the clumps in early spring, or sow seeds, which should germinate in three to four weeks.

VENIDIUM FASTUOSUM-Monarch Daisy. Annual. Start inside for earliest blooms. Germinates in about a month at 70 to 80 degrees F.

VERBASCUM-Mullein. Hardy biennial and perennial. Seeds germinate best at high temperatures and should be planted as soon as they ripen in mid-summer.

VERBENA. Tender perennial, usually treated as an annual. Seeds germinate in two to four weeks. Some varieties, such as *V. canadensis*, can be propagated by softwood cuttings or by dividing the plant.

VERONICA-Speedwell. Hardy perennial that can be easily grown from seed or from divisions of the clump in spring. Cuttings taken in early summer also root well.

VINCA MINOR-PERIWINKLE. MYR-TLE. Hardy perennial ground cover, easily propagated by divisions, cuttings, or layers. *V. major* is a tender perennial propagated the same way.

VIOLA-VIOLET. Annual and hardy perennial. Started easily from seeds or division of plants.

VIRGINIA BLUEBELL. See MER-TENSIA

WALLFLOWER. See CHEIRANTHUS

WANDERING JEW. See TRADES-CANTIA

WATERLILY. See NYMPHAEA

YARROW. See ACHILLEA

YUCCA. Perennials of varying hardiness. Seeds grow slowly, so it is best to propagate by separating offshoots from the parent plant in early spring. Root cuttings are also used.

ZEBRINA PENDULA. Cuttings taken at almost any time root easily.

Root them in small pots for easy transplanting.

ZINNIA. Annual. Seed germinates in a week. Those planted inside need special care, as they are very susceptible to damping-off diseases. Extra light and heat (75 to 85 degrees F.) help. So does having a sterile medium, and using fungicides.

ZOYSIA. Vigorous lawn grass, suitable where winters are not severe. Propagate by removing plugs of grass from existing beds, and plant them in new or established lawns.

Appendix

Greenhouses

Four Seasons Greenhouses
910 Route 110
Farmingdale, NY 11735

Lord and Burnham
Melville, NY 11746

National Greenhouse Co.
P.O. Box 100
Pana, IL 62557

Books

*Manual of Woody Landscape Plants**
by Dr. Michael A. Dirr
536 pages

Growing and Saving Vegetable Seeds
by Marc Rogers
Garden Way Publishing
114 pages

*Hortus III**
1,290 pages

*Plant Propagation: Principles and Practices**
by Hartman and Kester
Prentice Hall
662 pages

The Modern Nurseryman
by John Stanley and Alan Toogood

*Tissue Culture***
by Dr. R. A. Defossard

*Experiments in Plant Tissue Culture****
by Dodds and Roberts
Cambridge University Press

*Introduction to In Vitro Propagation***
by Donald Wetherell

Nursery Supply Houses

Walter E. Clark and Son
P.O. Box 754
Orange, CT 06477
grafting and pruning supplies

Geiger Corp.
Box 285
Harleysville, PA 19438
complete list of nursery and greenhouse supplies

A. M. Leonard
6665 Spiker Road
Piqua, OH 45356
also pruning, forestry, and grafting supplies, soil test kits

Orchard Equipment & Supply Co.
P.O. Box 146
Conway, MA 01341
also grafting supplies

Slater Supply Co.
143 Allen Blvd.
Farmingdale, NY 11735

Timm Enterprises Ltd.
P.O. Box 157
Oakville, Ontario, Canada
L6J 4Z5
horticultural equipment and machines

Tissue Culture Equipment and Media

Carolina Biological Supply Co.
2700 York Road
Burlington, NC 27215 or

Box 187
Gladstone, OR 97027

Scott Laboratories, Inc.
Fiskeville, RI 02823

Tree and Shrub Seed

Lawyer Nursery
950 Highway 200 West
Plains, MT 59859

F. W. Schumacher Co.
Horticulturists
Sandwich, MA 02563

Perennial and Herb Seed

W. Atlee Burpee Co.
5395 Burpee Bldg.
Warminster, PA 18974

Carter Seeds
475 Mar Vista Dr.
Vista, CA 92083
seeds for warm climates, catalog $1.00

Digiorgi Seed Co.
Council Bluffs, IA 51502

G. W. Park Seed Co.
S.C. Hwy 254
N. Greenwood, SC 29647

Scions for Grafting

Worcester County
Horticultural Society
30 Elm St.
Worcester, MA 01608

---

* Books are available from The American Nurseryman, 310 Michigan Ave., Suite 302, Chicago, IL 60604. Write for their prices and ordering information.

** Available from Dow Seeds Hawaii Ltd., Box 30144, Honolulu, HI 96820. (Price $48.70)

*** Books are available from Carolina Biological Supply Company, 2700 York Road, Burlington, NC 27215, or Box 187, Gladstone, OR 97027.

Suppliers of Seedlings (wholesale only)

Canales
Shelocta (Indiana County),
PA 15774

Hess Nurseries, Inc.
Box 326
Cedarville, NJ 08311

Western Maine Nurseries, Inc.
39 Elm St.
Fryeburg, ME 04037

Dwarf Rootstocks for Grafting

Grootendorst Nursery
Lakeside, MI 49116

Fruit Tree Seedlings for Grafting

Mt. Arbor Nurseries
Box 129
Shenandoah, IA 51601

Glossary

Agar. Jelly-like substance used in tissue culture.

Air layering. The starting of new plants asexually by developing roots on branches above ground that are still attached to the plant.

Annual plant. One that blooms, produces seeds, and dies in one year.

Asexual reproduction. The propagation of a plant by cuttings, division, layers, grafts, or other vegetative means, rather than by seeds. Used to produce plants that will be like their parent, and is sometimes called cloning.

Bare-rooted plant. Tree or shrub that is dug and transplanted without soil attached to its roots.

Basal. At the bottom or base of a plant. A basal shoot is a limb growing near the ground; a basal cut is made at the bottom of a cutting.

Bench. Popular term for a greenhouse shelf or table.

Bench grafting. Grafting that is done inside. Scions are attached to bare-rooted seedlings, pieces of roots, or potted seedlings.

Biennial plant. One with a life cycle of two growing seasons. It usually grows the first season, blooms and produces seeds the second season, and then dies.

Bottom heat. Heat, usually artificially supplied, beneath flats of seedlings, cuttings, or grafts, to induce more rapid rooting or growth.

Budding. A form of grafting in which a single bud is used rather than a branch scion.

Bud-graft. Same as budding.

Bud stick. The branch from which buds are taken.

Bulb. The fleshy root, with scales, of plants such as tulips, lilies, etc.

Bulbil. Tiny bulbs that form along the stem of certain plants such as tiger lilies and bladder ferns.

Bulblet. Small bulbs that develop around larger bulbs in the ground.

Callus. A fleshy tissue growth that forms on a plant while a wound is healing. On cuttings it sometimes, but not always, precedes the formation of roots.

Cambium. The actively growing layer of plant tissue which later turns into wood. It is the green layer just under the bark.

Chinese layering. Same as air layering.

Chip budding. Similar to regular budding, except that a chip of wood is included with the bud as it is inserted.

Cleft grafting. A method of grafting in which the scion is sharpened like a wedge and inserted in a split in the rootstock.

Clone. Plant that is started asexually. Especially one started by tissue culture.

Cold frame. An outside bed enclosed by a frame with a removable transparent cover, intended for growing plants with no artificial heat.

Compatibility. Scions and rootstocks that are able to grow together after grafting, to make a long-lasting union. Chemical and biological differences, weather, plant viruses, and other factors affect compatibility.

Compost. A rich porous soil made up of well-rotted organic matter. Excellent for enriching regular soil.

Compound layering. The laying down of a vine or long trailing branch and burying it in the soil at intervals to form new plants which root at the buried portions. Grapes, ivy, and other vines start easily this way, as do willows and poplars.

Corm. A fleshy root resembling a bulb, but solid. Corms, such as those from a gladiolus, can be cut into pieces to start new plants.

Cormels. Miniature corms that form around the parent corm, which can be removed and planted to start new plants.

Crown. The part of the plant above the trunk or stem that produces branches, leaves, and in most cases flowers and seeds. On herbaceous plants and spreading woody shrubs, the part of the plant that is just below the ground, from which the roots grow, to the part just above the ground from which the branches or leaves sprout.

Cultivar. A "named" variety of a plant that is different from others of the species. It usually, but not always, must be propagated by asexual means rather than by seeds.

Cutting. A piece of branch, root, or leaf that is separated from a plant, and used to create a new plant.

Desiccate. To lose water and dry out.

Dibble. A pointed tool used to make a hole in the soil for inserting a seed, plant, or bulb.

Dicotyledon. A plant that produces two seed leaves at germination.

Dominant. The controlling effect one plant has upon another plant, or a part of that plant. Also the ultimate dominating effect of an inherited characteristic on a seedling plant.

Dormancy. A resting period when the plant is not growing, and often not showing any signs of life, yet it is alive. Can also refer to seeds.

Dwarf stock. A rootstock upon which a plant is grafted for the purpose of creating a smaller tree than would grow naturally.

Embryo. The plantlet developing within a seed.

Eye. Another word, in plant terminology, for a bud.

Field grafting. Grafting on rootstock that is already growing in the place where it will be left to grow for a while.

Fungicide. A chemical used to control the growth of fungi.

Germination. The sprouting of seeds.

Girdling. The removal of a strip of bark all around a trunk of limb. It is used to induce rooting in air layers, or for putting on a patch bud.

Grafting. The transplanting of a portion of a branch of one plant to another for the purpose of forming a union so that they will grow as one plant.

Harden off. The process by which plants that have been started in an inside environment are gradually adjusted to the outdoors by controlling the temperature, nutrients, or other conditions.

Hardwood cutting. Cutting taken when the plant is dormant.

Heading back. Cutting back the top or terminal part of a plant, usually to induce side branching or root development. Also used to get rid of the aboveground part of seedling understock on a plant that has been bud-grafted or side-grafted.

Heel cutting. A cutting of new growth with a bit of old wood included at the base of the cutting. These form a heavier root system faster on some varieties.

Heel in. To plant temporarily, by burying the roots of plants in a trench, to keep them from drying out before they can be safely planted in their permanent location.

Herbaceous plant. One in which the part above the ground has a soft stem and does not become woody.

Herbicide. A chemical used for killing plants, usually weeds.

Humidity. The amount of water vapor in the atmosphere.

Hybrid. A new plant produced by cross pollination of two plants that are genetically different.

Incompatibility. See Compatibility.

Initial. Very first development of a plant.

Insecticide. A chemical for killing insects.

Juvenile. A young, immature plant that has not yet bloomed or fruited.

Latent bud. A bud that will not ordinarily develop further or start to grow, unless stimulated by pruning or in some other way.

Lateral bud. A side bud, as distinguished from a bottom or a terminal bud.

Lath. A frame or cover made of narrow slats to provide shade.

Layering. The process of bending over a lower limb, or spreading out the stem of a plant, and covering a portion of it with soil so it will root and form a new plant, which can later be severed from the parent.

Leaf cutting. A single leaf or a portion of a leaf that is cut to propagate a plant.

Line out. To plant young seedlings or cuttings in the open ground in a row, fairly close together, for later transplanting, potting, or selling.

Liner. A small grafted plant, rooted cutting, or seedling that is ready for transplanting into a pot, transplant bed, or nursery row.

Mallet cutting. A stem cutting, often of a vine, that includes a bit of older stem along with the current season's growth.

Mature plant. One that is old enough to bloom and produce seeds.

Mature wood. New growth that has hardened late in the season. The cuttings of some varieties of plants root better when taken from mature or partially mature wood.

Mist system. A fine mist of water sprayed continuously or intermittently over cuttings to keep them from drying out until they form roots.

Monocotyledon. A plant that produces only one seed leaf. Asparagus, most grains, grasses, palms, and similar plants are in this group.

Mound layering. Another name for stooling.

Mutation. A genetic change in a cell. Used to describe a part of a plant that changes quite suddenly, and starts to produce fruit, flowers, or leaves that are different from the rest of the plant.

Named variety. A variety of a plant that is in some way different from or superior to the ordinary species, which has been assigned a name. Such plants do not usually come true from seeds and must be propagated asexually. Also called a cultivar.

Node. The place where a leaf or bud comes out of a plant's stem.

Nurse graft. A rootstock that is used to support a scion long enough for it to grow roots of its own.

Offshoot. Stems coming from the base of a plant just under the ground. These often have roots of their own and can be removed from the parent plant and grown into new plants.

Partially mature wood. See Mature wood.

Patch bud. A method of budding in which a larger than normal piece of bark is included with the bud when it is placed on the rootstock.

Peat pellets. Small pellets of peat that swell up when watered. They may be used instead of pots and soil for starting seeds or rooting cuttings.

Perennial. A plant that lives for more than two years. Usually refers to herbaceous flowering plants, and may or may not be winter hardy.

Perennial plant. One that lives from year to year. The term usually refers to herbaceous flowering plants.

Petiole. The stalk of a leaf.

pH. The acidity-alkalinity condition of the soil. Most soils range from 4 to 8, and 7 is considered neutral. Ideal conditions range from about 4.5 for blueberries and azaleas to 7 for delphinium and lilacs. Most plants grow well at a pH that ranges from 5 1/2 to 7.

Photosynthesis. The process whereby the plant uses light to create carbohydrates from carbon dioxide, water, oxygen, and various minerals.

Pollen. The dust-like particles produced by the male stamens on a flower. Usually they are yellow or brown in color.

Pollination. The fertilization of the female ova of a plant by the transfer of pollen from the male portion of the same or a different flower, resulting in a seed. This process may be accomplished by insects, wind, or artificial means.

Rhizome. The fleshy roots of certain plants such as iris that serve as food storage organs.

Root cutting. Small pieces of roots that have been cut up for the purpose of starting new plants asexually. Certain plant varieties can be propagated easily in this manner.

Rooting chemical or **rooting compound.** A chemical in liquid or powder form that aids in the development and growth of roots on cuttings.

Runner. Vine-like growths from plants, such as strawberries, that produce new plants by natural layering.

Scaling a bulb. Removing the outer scales from a bulb and planting them to start new plants.

Scarification. A process used to break the dormancy of seeds that have hard shells. Soaking in warm water, acid treatment, filing the seed coats, and other methods are used to induce faster germination.

Scion (pronounced sigh-on). The branch or bud is grafted or bud-grafted onto a rootstock.

Scooping. A method used to cut out the base of hyacinths to encourage them to form small bulblets faster.

Seedling rootstock. A rootstock grown for grafting purposes that has been produced from seeds, as opposed to one grown from stooling or layering.

Selection. Choosing the best of a group of seedlings or wild plants for propagation or breeding.

Serpentine layers. Another term for compound layers.

Shield budding. The same as T-budding.

Side veneer grafting. A graft place on the side of a plant. This method is usually used on potted evergreens, and is done in greenhouses.

Simple layering. Regular layering.

Softwood cutting. A cutting made early in the season, from new growth.

Spore. A reproductive unit, consisting of one or more cells, capable of producing a new plant, either directly or indirectly.

Sport. Same as mutation.

Standard. A standard or standard-sized tree is one that grows to normal size, as opposed to one that is grafted on dwarf or semi-dwarf roots. Also, the upright form in which certain spreading plants, such as roses and fuchsias are sometimes trained.

Stock. Rootstock.

Stooling. A method of propagation in which soil is piled over the base of an existing plant encouraging roots to form at the base of the branches, in the soil.

Stratification. The storage of seeds at a certain temperature and humidity for a period of time, for the purpose of helping germination. Seeds planted outdoors in the fall as soon as they are ripe often get these conditions naturally.

Succulent. A plant that is capable of storing large amounts of moisture for long periods of time.

Sucker. A shoot or branch growing from the base of a plant, either above or below ground level. Also, plants growing from roots of the parent plant, sometimes emerging quite a distance from the parent itself.

Systemics. Chemicals absorbed by a plant that ultimately permeate the entire plant. They may be absorbed through the roots, or by spraying the above-ground portions.

Tap root. The main root of a plant.

T-budding. The most common kind of bud grafting, sometimes called shield budding, that consists of placing a bud within a T cut in the bark of the stock.

Terminal bud. The end bud on the top of a plant or the tip of a stem.

Tip layering. The layering of plants in which only the tip end of the plant is buried in the soil, where it roots. Black raspberries root naturally in this way, for example.

Tissue culture. The asexual propagation of plants in a laboratory by culture and careful control of cell growth. Sometimes called "test-tube cloning."

Top working. The grafting of a sizable tree, in which several limbs are changed to a different variety or varieties.

Transpiration. The process of giving off water vapor by a plant, mostly through the leaves.

Trench layering. A layering method whereby a whole limb or an entire small tree is buried in a trench, with the base portion of all upright growth buried in soil, as soon as the growth is large enough. It is used to establish a grafted tree on its own roots, or to obtain several new plants from the parent.

Tuber. A fleshy root which is a food storage organ. Some develop annually, as on dahlias, and others are perennial, as on certain begonias.

Turgid. This is used to describe a plant that is adequately supplied with water. The cuttings from turgid plants root better than non-turgid plants as a rule, and scions or buds from them are better as grafts.

Union. The place on a tree or plant where the rootstock and scion have grown together. In certain trees this juncture is quite noticeable, even after the tree is many years old.

Vegetative. The growing, as opposed to the flowering, part of a plant. Often used to describe asexual methods of propagating plants as opposed to starting them from seed.

Veneer grafting. A form of side grafting in which a chip of wood attached to a small branch is grafted on the side of a seedling. This method is commonly used on potted evergreens and done in greenhouses.

Viability. The ability of a seed to germinate and grow.

Water stress. A condition whereby a plant is losing water faster than it can absorb it.

Wedge graft. Another name for a cleft graft.

Whip and tongue grafting. A grafting process in which the scion and rootstock are locked together tighter than in ordinary grafting, often used in bench grafting.

Woody plant. A plant with woody tissue, as opposed to those with soft, herbaceous stems.

Wounding. The cutting away of a piece of bark near the base of a cutting or layer to expose the wood tissue and stimulate faster rooting at that point. Rooting chemicals are often applied to the wounded parts.

Index

The numbers in boldface refer to illustrations.